Cohomology of Quotients in Symplectic and Algebraic Geometry

by

Frances Clare Kirwan

Mathematical Notes 31

Princeton University Press

Princeton, New Jersey

1984

Printed in the United States of America by
Princeton University Press, 41 William Street,
Princeton, New Jersey 08540

ISBN 0-691-08370-3

The Princeton Mathematical Notes are edited by William Browder,
Robert Langlands, John Milnor, and Elias M. Stein

Library of Congress Cataloging in Publication Data
will be found on the last printed page of this book

Contents

§1. Introduction

The aim of these notes is to develop a general procedure for computing the rational cohomology of quotients of group actions in algebraic geometry. The main results were announced in [Ki].

We shall consider linear actions of complex reductive groups on nonsingular complex projective varieties. To any such action there is associated a projective "quotient" variety defined by Mumford in [M]. This quotient variety does not coincide with the ordinary topological quotient of the action. For example, consider the action of $SL(2)$ on complex projective space $\mathbf{P}_n$ where $\mathbf{P}_n$ is identified with the space of binary forms of degree n, or equivalently of unordered sets of n points on the projective line $\mathbf{P}_1$. The orbit where all n points coincide is contained in the closure of every other orbit and hence the topological quotient cannot possibly be given the structure of a projective variety. To obtain a quotient which is a variety such "bad" orbits have to be left out.

The quotient variety can be described as follows. Suppose that X is a projective variety embedded in some complex projective space $\mathbf{P}_n$ and that G is a complex reductive group acting on X via a homomorphism $\phi : G \to GL(n+1)$. If $A(X)$ denotes the graded coordinate ring of X, then the invariant subring $A(X)^G$ is a finitely-generated graded ring: let M be its associated projective variety. The inclusion of $A(X)^G$ in $A(X)$ induces a G-invariant surjective morphism ψ from an open subset X^{ss} of X to M.

(The points of X^{ss} are called semistable for the action). There is an open subset M' of M which is an orbit space for the action of G on its inverse image under ψ, in the sense that each fibre is a single orbit of G.

So we have two 'quotients' M and M' associated to the action of G on X. Our main purpose here is to find a procedure for calculating the cohomology, or at least the Betti numbers, of these in the good cases when they coincide. This happens precisely when M is topologically the ordinary quotient X^{ss}/G. In fact we make the slightly stronger requirement that the stabiliser in G of every semistable point of X should be finite; this is equivalent to requiring that every semistable point should be (properly) stable. Under these conditions an explicit formula is obtained for the Betti numbers of the quotient M (see theorem 8.12.) This formula involves the cohomology of X and certain linear sections of X, together with the classifying spaces of G and certain reductive subgroups of G.

For example, consider again the action of $SL(2)$ on binary forms of degree n. Then good cases occur when n is odd, and one finds that the non-zero Betti numbers of the quotient M are given by

$$\dim H^{2j}(M;Q) = \left[1 + \frac{1}{2}\min(j,n-3-j)\right]$$

for $0 \leq j \leq n - 3$.

Our approach to the problem follows the method used by Atiyah and Bott to calculate the cohomology of moduli spaces of vector bundles over Riemann surfaces [A & B]. It consists in finding a canonical stratification[1] of X associated to the action of G whose unique open stratum coincides with the set X^{ss} of semistable points provided $X^{ss} \neq \emptyset$. There are then Morse-type inequalities relating the Betti numbers of X to those of X^{ss} and the other strata; and since the stratification is G-invariant there also exist equivariant Morse inequalities which turn out in fact to be equalities.

Recall that the rational equivariant cohomology $H_G^*(Y;Q)$ of a space Y acted on by G is defined to be $H^*(Y_G;Q)$ where $Y_G = Y \times_G EG$ and $EG \to BG$ is the universal classifying bundle for G. When the rational equivariant Morse inequalities of a stratification are equalities they can be stated in the form

1.1 $$\dim H_G^n(Y;Q) = \sum_S \dim H_G^{n-\lambda(S)}(S;Q)$$

for each $n \geq 0$, where the sum runs over all the strata S of the stratification and $\lambda(S)$ is the codimension of S in X (see [A & B] §1). Moreover using the assumption that every point of X^{ss} has finite stabiliser in G we can show that

$$H_G^*(X^{ss};Q) = H^*(X^{ss}/G;Q) = H^*(M;Q).$$

Hence the Morse equalities will give us formulae for the Betti numbers of M in terms of the rational equivariant cohomology of X and of the other strata.

The Morse inequalities are obtained by building up X from the strata and using the Thom-Gysin sequences of rational equivariant cohomology that occur every time a stratum is added. Of course any coefficient field may be used instead of Q, but then the Morse inqualities are not necessarily equalities and the cohomology of the quotient M may not be isomorphic to the equivariant cohomology of X^{ss}. So information about the torsion of M can only be obtained in special cases.

As in [A & B] there are two different approaches to the problem of defining a suitable stratification. One approach is purely algebraic, and leads to a definition of a stratification given a linear reductive group action on a projective variety defined over any algebraically closed field. This method will be developed in Part II. It is based on work of Kempf (see [K] and [Hes] and [K&N]). The paper [Ne] by Ness has very close links with much of what is covered here and in Part I, although our results were arrived at independently.

The alternative approach is based on Morse theory and symplectic geometry, and will be developed in Part I. The idea is to associate a certain function f in a canonical way to the action of G on X and use it to define a "Morse stratification" of X. The stratum to which any point of X belongs is determined by the limit of its path of steepest descent for the Kähler metric under the function f.

The advantages of this approach are that it is conceptually simpler and that it can be applied to compact Kähler manifolds as well as to nonsingular projective varieties. More generally still it enables us to calculate the rational cohomology of the "symplectic quotient", when it exists, of any symplectic manifold by the action of a compact Lie group.

The function to which Atiyah and Bott apply the methods of Morse theory in their special case (where the group and the space are both infinite-dimensional) is the Yang-Mills functional. As pointed out in [A & B] the latter can be described in terms of symplectic geometry as the norm-square of the moment map. But in this form it makes sense in our situation.

Recall that a symplectic manifold X is a smooth manifold equipped with a nondegenerate closed 2-form ω, and a compact Lie group K acts symplectically on the manifold if it acts smoothly and preserves ω. Associated to such an action one has the concept of a moment map $\mu: X \to \mathbf{k}^*$ where $\mathbf{k}^*$ is the dual of the Lie algebra of K. For example when $SO(3)$ acts on the cotangent bundle $T^*\mathbf{R}^3$ ("phase space") the moment map can be identified with angular momentum. The existence of a moment map is guaranteed by conditions such as the semisimplicity of K or the vanishing of $H^1(X;\mathbf{Q})$.

Consider again a reductive group G acting on a nonsingular complex projective variety $X \subseteq \mathbf{P}_n$ via a homomorphism $\phi: G \to GL(n+1)$. Since G is reductive it is the complexification of a maximal subgroup K. We may assume that K acts unitarily on C^{n+1} and so preserves the standard Kähler structure on $\mathbf{P}_n$. This Kähler structure makes X into a symplectic manifold on which K acts symplectically. (It also gives a natural choice of Riemannian metric on X). There is a moment map $\mu: X \to \mathbf{k}^*$ associated to this action which can be described explicitly (see 2.7). If we fix an invariant inner product on the Lie algebra of K then the norm-square of the moment

μ provides us with a K-invariant Morse function f on X.

Unfortunately this Morse function is not nondegenerate in the sense of Bott, so the results of Morse theory cannot be applied to it directly. To avoid this problem one can use the approach of Part II to define the stratification algebraically and prove that it has all the properties one wants, showing later that it is in fact in a natural sense the Morse stratification for f. On the other hand if one is prepared to do a little local analysis one can extend the arguments of Morse theory to degenerate functions which are reasonably well-behaved. It will be shown that the norm-square of the moment map is well-behaved in this sense: this is true when X is any symplectic manifold acted on by a compact group K.

More precisely, we shall see that the set of critical points for the function $f = \|\mu\|^2$ is a finite disjoint union of closed subsets $\{C_\beta \mid \beta \in \mathbf{B}\}$, along each of which f is minimally degenerate in the following sense. A locally closed submanifold Σ_β containing C_β with orientable normal bundle in X is a minimising manifold for f along C_β if

(i) the restriction of f to Σ_β achieves its minimum value exactly on C_β, and

(ii) the tangent space to Σ_β at any point x of C_β is maximal among subspaces of T_xX on which the Hessian $H_x(f)$ of f is positive semi-definite.

If a minimising manifold Σ_β exists then f is called minimally degenerate along C_β.

In the appendix it is shown that these conditions imply that f induces a smooth stratification $\{S_\beta \mid \beta \in \mathbf{B}\}$ of X such that a point lies in the stratum S_β if its path of steepest descent for f has a limit point in the critical subset C_β. (For this X must be given a suitable metric; when X is a Kåhler manifold and $f = |\mu|^2$ the Kåhler metric will do). The stratum S_β then coincides with Σ_β near C_β. The proof is not hard but involves some analysis of differential equations near critical points.

As has already been mentioned, it turns out that the unique open stratum of this stratification coincides with the set X^{ss} of semistable points of X; and that in the good cases its G-equivariant rational cohomology is isomorphic to the ordinary rational cohomology of the quotient variety M, which is what we were after. Moreover the stratification is G-invariant and equivariantly perfect over the rationals, in the sense that its equivariant Morse inequalities are in fact equalities. Thus the formula 1.1 can be used to calculate the Betti numbers of M in terms of the equivariant cohomology of X itself and of the non-semistable strata.

In order that this formula should be useful it is necessary to investigate the non-semistable strata. It turns out that the equivariant cohomology of these can be calculated inductively. In fact each stratum S_β has the form

1.2 $$S_\beta \cong G \times_{P_\beta} Y_\beta^{ss}$$

where Y_β^{ss} is a locally-closed nonsingular subvariety of X and P_β is a parabolic subgroup of G (see theorem 6.18). This implies that the G-equivariant cohomology of S_β is isomorphic to the P_β-equivariant cohomology of Y_β^{ss}. Moreover there is a linear action of a maximal reductive subgroup of P_β on a proper nonsingular closed subvariety Z_β of X such that Y_β^{ss} retracts equivariantly onto the subset Z_β^{ss} of semistable points for this action. It follows that $H^*_{P_\beta}(Y_\beta^{ss};Q)$ is isomorphic to the rational equivariant cohomology of Z_β^{ss} with respect to this reductive subgroup. By induction we may assume that this is known.

It now remains to consider the equivariant cohomology $H^*_G(X;Q)$ of X itself. We shall assume for convenience that G is connected. Then one can show (see proposition 5.8) that the spectral sequence of the fibration

$$X_G = X \times_G EG \to BG$$

degenerates over the rationals (where $EG \to BG$ is the universal classifying bundle for G). This means that the equivariant cohomology of X is isomorphic to the tensor product $H^*(X;Q) \otimes H^*(BG;Q)$ of the cohomology of X and of BG. (It is easy to deduce from this what happens for disconnected groups. For if G has identity component Γ, then $H^*_G(X;Q)$ is the invariant part of $H^*_\Gamma(X;Q)$ under the action of the finite group G/Γ).

Thus the formula for $H^*_G(X^{ss};Q)$ in terms of the equivariant cohomology of X and of the non-semistable strata gives us an <u>inductive procedure</u> for calculating $H^*_G(X^{ss};Q)$. This leads to an <u>explicit formula for</u> $H^*_G(X^{ss};Q)$, and hence also in good cases for <u>the Betti numbers of the quotient variety</u> M. This formula involves the cohomology of X and certain nonsingular subvarieties of X together with the cohomology of the classifying spaces of G and various reductive subgroups of G (see theorem 8.12).

The stratification induced by the norm-square of the moment map has also been studied by Ness in [Ne]. Moreover related research on Betti numbers of quotients by C^* and $SL(2,C)$ actions has been done independently by Bialynicki-Birula and Sommese. In fact in their paper [B-B & S] they consider quotients of many different open subsets by G, not just X^{ss}, and completely classify those subsets for which quotients exist.

When X is merely a compact symplectic manifold acted on by a compact group K the function $f = \|\mu\|^2$ still induces a smooth stratification of X, although most of the structure of the Morse strata is lost. The loss of structure is to be expected because the stratification depends on choosing a K-invariant Riemannian metric on X and there is no longer a natural choice given by the real part of the K-invariant Kähler metric. So we concentrate on the <u>critical subsets</u> C_β instead (which are not necessarily submanifolds of X).

In fact the form in which the Morse inequalities are usually stated is that in which the cohomology of each Morse stratum S_β is replaced by that of its critical subset C_β. This replacement is allowable because the inclusion of C_β in S_β is an equivalence of both equivariant and ordinary (Čech) cohomology. These critical subsets C_β are independent of the choice of metric. They have the following description in terms of minimal sets for smaller manifolds which is analogous to 1.2. For each $\beta \in B$ there is a symplectic submanifold Z_β of X acted on by a compact subgroup $\text{Stab}\beta$ of G and a moment map μ_β for this action such that

$$C_\beta \cong K \times_{Stab\beta} \mu_\beta^{-1}(0).$$

Since f is equivalently perfect and

$$H_K^*(X;Q) \cong H^*(X;Q) \otimes H^*(BK;Q),$$

we obtain an inductive procedure for calculating the dimensions of the equivariant cohomology groups $H_K^n(\mu^{-1}(0);Q)$ of the minimum critical set $\mu^{-1}(0)$ for f.

The reason why $H_K^*(\mu^{-1}(0);Q)$ is interesting is that when a 'symplectic quotient' of the action of K on X exists then its rational cohomology is isomorphic to $H_K^*(\mu^{-1}(0);Q)$. In order that the symplectic quotient should exist in a reasonable sense one has to assume that there is a moment map μ: $X \to \mathbf{k}^*$ and that the stabiliser in K of every $x \in \mu^{-1}(0)$ is finite. Then one finds that $\mu^{-1}(0)$ is a submanifold of X, and that the Kähler form ω on X induces a symplectic structure on the topological quotient $\mu^{-1}(0)/K$ which is a manifold except for singularities due to the presence of finite isotropy groups. With this structure $\mu^{-1}(0)/K$ is the natural symplectic quotient (or Marsden-Weinstein reduction) of X by K. Because of the assumption on stabilisers its rational cohomology is isomorphic to $H_K^*(\mu^{-1}(0);Q)$.

As we have already seen, the link between algebraic and symplectic geometry is through Kähler geometry. Except for the connection with semistability and invariant theory the results for projective varieties hold when X is any compact Kähler manifold acted on by a complex group G, provided that G is the complexification of a maximal compact subgroup K which preserves the Kähler structure on X and that there exists a moment map μ: $X \to \mathbf{k}^*$. We obtain an equivariantly perfect stratification of X such that each stratum is a locally-closed complex submanifold of X and can be decomposed in a form analogous to that described at 1.2. Moreover it turns out that if the symplectic quotient $\mu^{-1}(0)/K$ exists then it can be identified with the quotient of the minimum stratum X^{min} by the complex group G. Because of this it can be given the structure of a compact Kåhler manifold, except for singularities caused by finite isotropy groups. So $X^{min}/G = \mu^{-1}(0)/K$ is a natural Kåhler quotient of X by G, and its Betti numbers can be calculated by the method already described.

In particular in the case of a linear action on a projective variety the quotient variety M obtained from invariant theory coincides topologically with the quotient $\mu^{-1}(0)/K$; in fact this is true in all cases, not only "good" ones.

The set $\mathbf{B}$ which indexes the critical subsets C_β and also the stratification can be identified with a finite set of orbits of the adjoint representation of K on its Lie algebra **k**. Each orbit in $\mathbf{B}$ is the image under the moment map $\mu: X \to \mathbf{k}^* \cong \mathbf{k}$ of the critical subset which it indexes. If a choice is made of a positive Weyl chamber $\mathbf{t}_+$ in the Lie algebra of some maximal torus of K then each adjoint orbit intersects $\mathbf{t}_+$ in a unique point, so $\mathbf{B}$ can be regarded alternatively as a finite set of points in $\mathbf{t}_+$. When $X \subseteq \mathbf{P}_n$ is a projective variety on which K acts linearly via a homomorphism $\phi: K \to GL(n+1)$ these points can be described in terms of the weights of the representation of K given by ϕ as follows: a point of $\mathbf{t}_+$ lies in $\mathbf{B}$ if it is the closest point to the origin of the convex hull of a nonempty set of these weights. (Recall that there is a fixed invariant inner product on $\mathbf{k}$ which is used to identify $\mathbf{k}$ with $\mathbf{k}^*$). This is true also in the general symplectic case if the definition of a 'weight' is extended appropriately.

In terms of this last description if $\beta \in \mathbf{B}$ then the submanifold Z_β of X which appeared in the inductive description of the critical subset C_β and of the stratum S_β is the union of certain components of the fixed point set of the subtorus of K generated by β. The subgroup Stabβ is the stabiliser of β under the adjoint action of K on its Lie algebra $\mathbf{k}$ and in the Kähler case the complexification of Stabβ is a maximal reductive subgroup of the parabolic subgroup P_β.

The function $f = \|\mu\|^2$ is not unique in possessing the properties described above. The same arguments work for any convex function of the moment map (cf. [A & B] §§8 and 12).

Finally it should be noted that the assumption of compactness is not essential (see §9). There are interesting examples of quasi-projective varieties and noncompact symplectic manifolds to which the same sort of analysis can be applied by taking a little extra care. These include the original examples of symplectic manifolds, viz. cotangent bundles.

The layout of the first part is as follows. §§2-5 are concerned with any symplectic action of a compact group K on a compact symplectic manifold X. In §2 we introduce the moment map μ, giving particular emphasis to the case when a compact group acts linearly on a nonsingular complex projective variety. We then describe the Morse stratification associated to a non-degenerate Morse function, and discuss how the ideas of Morse theory might be applied to the function $f = \|\mu\|^2$ even though it is degenerate. In §3 we describe the set of critical points for f as a finite disjoint union of closed subsets $\{C_\beta \mid \beta \in \mathbf{B}\}$. It is then shown in §4 that f is minimally degenerate along each critical subset C_β. This implies that there are Morse inequalities relating the Betti numbers of the symplectic manifold X to those of the subsets C_β; the proof of this fact is left to the appendix. In §5 these Morse inequalities are shown to be equalities for rational equivariant cohomology (see theorem 5.4). Inductive and explicit formulae are obtained for the dimensions of the cohomology groups $H^n_K(\mu^{-1}(0);Q)$, and it is shown that these coincide with the Betti numbers of the symplectic quotient $\mu^{-1}(0)/K$ when this exists.

The next two sections study the case when X is a Kähler manifold so that there is a natural choice of metric on X. In §6 we see that the function $f = |\mu|^2$ induces a Morse stratification $\{S_\beta \mid \beta \in \mathbf{B}\}$ with respect to this metric such that the strata S_β are locally-closed complex submanifolds of X and are invariant under the action of the complex group G. It is also shown that the strata S_β have the structure described at 1.2 above. The cohomological formulae of §5 are interpreted in the Kähler case, and there is a brief discussion of how the stratification is affected if the choices of moment map and of inner product on the Lie algebra $\mathbf{k}$ are altered. In §7 we see that if a symplectic quotient exists for the action of K on X

then it has a natural Kåhler structure and can be regarded as a 'Kåhler quotient' of the action of G on X.

Then in §8 we consider the case when G is a complex reductive group acting linearly on X which is a nonsingular complex projective variety. It is shown that the open subset X^{ss} of semistable points for the action coincides with the minimum stratum of the Morse stratification, so that §5 gives us an inductive formula for its rational equivariant cohomology. In 'good' cases when the stabiliser of every semistable point is finite we deduce that the projective quotient variety defined in geometric invariant theory coincides with the symplectic quotient $\mu^{-1}(0)/K$. Our original aim is then achieved by interpreting the formulae of §5 to give formulae for the Betti numbers of this quotient variety (see theorem 8.12).

§9 contains some remarks on how to loosen the requirement of compactness. Examples are given of formulae obtained by looking at the symplectic actions on cotangent bundles induced by arbitrary actions of compact groups on manifolds.

Part II gives an algebraic approach to the same problem. It is shown in §§12 and 13 that if k is <u>any algebraically closed field</u> and G is a reductive group acting linearly on a projective variety X defined over k then a stratification $\{S_\beta \mid \beta \in B\}$ of X can be defined which coincides with the stratification defined in Part I when k= C. The strata S_β are all G-invariant subvarieties of X. Moreover if X is nonsingular then so are the strata S_β, and they have the structure described at 1.2. This algebraic definition of the stratification relies heavily on work of Kempf (as expounded in [He]).

The fact that such stratifications exist when k is the algebraic closure of a finite field provides an alternative method for obtaining the formulae found in Part I for the Betti numbers of quotients of nonsingular complex projective varieties. For this one has to count points in quotients defined over finite fields, and then apply the Weil conjectures (see §15). This is the method used by Harder and Narasimhan in [H&N] to obtain the formulae later rederived by Atiyah and Bott for the Betti numbers of moduli spaces of vector bundles over a Riemann surface.

It is shown in §14 how the formulae for Betti numbers can be refined to give Hodge numbers as well. As an immediate corollary we have that if the Hodge numbers $h^{p,q}$ of the variety X vanish when $p \neq q$ then the same is true of the quotient variety.

In the final section some detailed examples are given of the stratifications and of calculating the rational cohomology of the quotients. One example studied is that of products of Grassmannians acted on by general linear groups. It will be shown in a future paper (see [Ki3]) that this can be used to give an alternative derivation of the formulae of [A & B] for the cohomology of moduli spaces of vector bundles over Riemann surfaces. This alternative derivation uses finite-dimensional group actions whereas in [A & B] the groups and spaces are all infinite-dimensional.

The formulae for Betti numbers obtained in this monograph depend upon the restrictive assumption that the stabiliser of every semi-stable point is finite. This assumption implies in particular that the quotient variety has only the minor singularities due to the existence of finite isotropy groups, whereas in general the quotient has more serious singularities. However provided that X^s is not empty one can obtain interesting information even when there are semistable points which are not stable. In fact there is a canonical way to blow up X along a sequence of nonsingular subvarieties to obtain a projective variety $\tilde{X}$ with a linear action of G for which every semistable point is stable. Then the geometric invariant theory quotient of $\tilde{X}$ (which has only minor singularities) can be regarded as an approximate desingularisation of the quotient of X, and there is a formula for its Betti numbers similar to that of theorem 8.12 (see [Ki2]).

Finally there are some differences of notation and also some inaccuracies in the announcement of these results in [Ki]. One mistake is that the theorem as it stands is only valid when G is connected, because remark (1) is only true in this case. Another is that in (d) of the proposition it is only the reductive part $\mathrm{Stab}\beta$ of the parabolic subgroup P_β which acts on Z_β not the whole of P_β. Furthermore the last sentence might be taken to imply that the geometric invariant theory quotient of a product of Grassmannians is torsion-free. This is not true since the projective linear groups $PGL(m,\mathbf{C})$ have torsion.

I would like to thank all those who gave me help and advice, including Michael Penington, Simon Donaldson, Michael Murray, John Roe, Graeme Segal and the referee, and to thank Linda Ness for sending me her results. I also thank Laura Schlesinger for her excellent typing, and the Science and Engineering Research Council of Great Britain for a grant which supported me during the course of my research. Above all I wish to acknowledge my great debt to my supervisor Michael Atiyah, to whom most of the basic ideas of these notes are due.

Footnote

[1] It has been pointed out by the referee that the term "stratification" is usually reserved for a decomposition which is topologically locally trivial in a neighbourhood of each stratum (Whitney stratifications, for example). The stratifications in these notes are not required to satisfy this property (see definition 2.11 below). Perhaps they should be more properly called "manifold decompositions".

Notation

For the convenience of the reader there follows a list of some of the notation used throughout.

Part 1.

K	compact Lie group with Lie algebra $\mathbf{k}$
G	complexification of K with Lie algebra $\mathbf{g}$
X	compact symplectic manifold acted on by K (later a compact Kåhler manifold acted on by G)
ω	symplectic or Kåhler form on X
$\mu: X \to \mathbf{k}^*$	moment map (see definition 2.3)
$\cdot$	invariant inner product on $\mathbf{k}$ used to identify $\mathbf{k}^*$ with $\mathbf{k}$ throughout
$f = \|\mu\|^2$	norm-square of μ with respect to inner product on $\mathbf{k}$
$\{S_\beta \mid \beta \in \mathbf{B}\}$	stratification of X induced by f
$EG \to BG$	universal classifying bundle of G
H_G^*	equivariant cohomology (see 2.15)
P_t	ordinary Poincaré series
P_t^G	equivariant Poincaré series
$x \to \beta_x$	vector field on X induced by $\beta \in \mathbf{k}$
T	maximal torus of K with Lie algebra $\mathbf{t}$
$\mathbf{t}_+$	positive Weyl chamber
$\mu_T: X \to \mathbf{t}^*$	moment map for T (3.3)

T_β	subtorus of T generated by $\beta \in \mathbf{t}$ (3.7)
μ_β	projection of μ along β (3.7)
Z_β	set of fixed points of T_β (or critical points of μ_β) on which μ_β takes the value $\|\beta\|^2$ (3.10)
$\mathbf{B}$	indexing set; finite subset of $\mathbf{t}_+$ (3.13)
$C_\beta = K(Z_\beta \cap \mu^{-1}(\beta))$	critical set corresponding to $\beta \in \mathbf{B}$ (3.14)
J	almost complex structure on X (4.1)
$Y_\beta \supseteq Z_\beta$	Morse stratum (or 'stable manifold') of Z_β with respect to μ_β (4.6)
$p_\beta : Y_\beta \to Z_\beta$	retraction (6.3)
Stabβ	stabiliser of β under adjoint action of K on $\mathbf{k}$ (4.8)
$S_\beta \supseteq C_\beta$	Morse stratum (or 'stable manifold') of C_β with respect to f
KY_β	minimising manifold for f along C_β (see 10.1); coincides with S_β locally along C_β
$H_x(f)$	Hessian of f at x
$Z_{\beta,m}$	union of certain components of Z_β (4.19)
$C_{\beta,m}$	union of corresponding components of C_β
$d(\beta,m)$	index of $H_x(f)$ for $x \in C_{\beta,m}$ (4.20)
$\underline{\beta}$	β-sequence (5.11)
Z_β^{min}	minimum Morse stratum for $\|\mu-\beta\|^2$ on Z_β (6.3)
$Y_\beta^{min} = p_\beta^{-1}(Z_\beta^{min})$	open subset of Y_β

$S_\beta = GY_\beta^{min}$	in Kåhler case (6.18)
B	Borel subgroup of G (6.8)
$P_\beta = B\ \mathrm{Stab}\beta$	parabolic subgroup of G (6.9)

Part II.

X	nonsingular projective variety over an algebraically closed field k
G	reductive group acting linearly on X
X^s	stable points of X
X^{ss}	semistable points of X
$m(x^*,\lambda)$	measure of instability (2.1)
$M(G)$	set of virtual one-parameter subgroups of G (2.4)
q	norm on $M(G)$ (2.4)
$\Lambda_G(x)$	see 2.5
Z_β^{ss}	semistable points of Z_β under an appropriate action of $\mathrm{Stab}\beta$ (2.20 and 2.21)
$Y_\beta^{ss} = p_\beta^{-1}(Z_\beta^{ss})$	open subset of Y_β

Part I. The symplectic approach.

§2. The moment map

This section introduces the concept of a moment map associated to a compact group action on a symplectic manifold. Special emphasis will be given to the examples of most interest to us, which are linear actions on nonsingular complex projective varieties. A precise formula is given at 2.7 for the moment map in these cases.

The moment map will be used to define a real-valued function on the symplectic manifold concerned. We shall conclude the section by considering how the ideas of Morse theory might be applied to this function, in spite of the fact that it is not a nondegenerate Morse function.

A symplectic manifold is a smooth manifold X equipped with a nondegenerate closed 2-form ω. A compact Lie group K is said to act symplectically on X if K acts smoothly and $k^*\omega = \omega$ for every $k \in K$. We shall assume throughout that every compact group action on a symplectic manifold is symplectic unless specified otherwise.

Any Kähler manifold X can be given the structure of a symplectic manifold by taking for ω the Kähler form on X, which is the imaginary part of a hermitian metric η on X. If K is any compact Lie group acting on X then the average $\int_K k^*\eta$ of η over K is a Kähler metric whose imaginary part is a K-invariant symplectic form on X.

The special case which will be of most interest to us is the following.

2.1. Example: linear actions on complex projective varieties

Let X be a nonsingular subvariety of some complex projective space $\mathbf{P}_n$ and suppose that a compact Lie group K acts on $\mathbf{P}_n$ via a homomorphism $\phi: K \to GL(n+1)$. (We assume that $GL(n+1)$ acts on $\mathbf{P}_n$ as left multiplication of column vectors by matrices). By conjugating ϕ with a suitable element of $GL(n+1)$ we may assume that $\phi(K)$ is contained in the unitary group $U(n+1)$. The restriction of the Fubini-Study metric on $\mathbf{P}_n$ then gives X a Kähler structure which is preserved by K.

2.2. Example: configurations of points on the complex sphere

A particular case of 2.1 which will be used throughout to illustrate definitions and results is that of the diagonal action of SU(2) on the spaces $(\mathbf{P}_1)^n$ of sequences of points on the complex sphere. $(\mathbf{P}_1)^n$ is embedded in $\mathbf{P}_{2^n-1}$ by the Segre embedding. Alternatively one can consider the action of SU(2) on the space of unordered sets of n points in $\mathbf{P}_1$, which can be identified with $\mathbf{P}_n$.

Let $\mathbf{k}$ be the Lie algebra of K and let $\mathbf{k}^*$ be its dual. Then a moment map (or momentum mapping) for the action of K on X is a map $\mu: X \to \mathbf{k}^*$ which is K-equivariant with respect to the given action of K on X and the co-adjoint action Ad^* of K on $\mathbf{k}^*$ and satisfies the following condition.

2.3. <u>For every</u> $a \in \mathbf{k}$ <u>the composition of</u> $d\mu\colon TX \to \mathbf{k}^*$ <u>with evaluation at</u> a <u>defines a 1-form on</u> X. <u>This 1-form is required to correspond under the duality defined by</u> ω <u>to the vector field</u> $x \to a_x$ <u>on</u> X <u>induced by</u> a. That is, for all $x \in X$ and $\xi \in T_xX$

$$d\mu(x)(\xi).a = \omega_x(\xi, a_x)$$

where . denotes the natural pairing of $\mathbf{k}^*$ and $\mathbf{k}$. In other words the component of μ along a is a Hamiltonian function for the vector field on X defined by a.

μ is determined up to an additive constant by 2.3. When K is semisimple μ is determined completely, since the only point of $\mathbf{k}^*$ fixed by the co-adjoint action is 0. If on the other hand K is a torus the addition of a constant to μ does not affect its equivariance because K acts trivially on $\mathbf{k}^*$. However if a moment map μ exists we can always make a canonical choice of μ by requiring that the integral of μ over the manifold X (with the highest exterior power of ω as volume form) should vanish.

By a theorem of Marsden and Weinstein (see [M & W]) a moment map $\mu\colon X \to \mathbf{k}^*$ always exists (and is unique) when K is semisimple. In addition if $H^1(X;\mathbf{Q}) = 0$ then a moment map always exists when K is a torus. For the adjoint action of a torus on its Lie algebra is trivial, so by 2.3 we just have to solve the differential equations

$$d\mu(x)(\xi).a = \omega_x(a_x, \xi)$$

for each a in some basis of the Lie algebra $\mathbf{k}$. This is possible if $H^1(X;Q) = 0$ since $d\omega = 0$.

A compact Lie group is the product of a semisimple group and a torus, at least modulo finite central extensions. Moreover if $K_1 \to K_2$ is a finite central extension then a moment map for K_1 is the same as a moment map for K_2. It follows that a moment map always exists when $H^1(X;Q) = 0$.

It is easy to see that when K acts on $X \subseteq P_n$ via a homomorphism $\phi: K \to U(n+1)$ a moment map always exists. It is sufficient to prove existence when $U(n+1)$ acts on $\mathbf{P}_n$ since if μ is a moment map for this action then the composition

2.4. $$X \to \mathbf{P}_n \xrightarrow{\mu} \mathbf{u}(n+1)^* \xrightarrow{\phi^*} \mathbf{k}^*$$

is a moment map for the action of K on X. But we have

2.5. Lemma. Let $x^* = (x_0,\dots,x_n)$ be any nonzero vector of $\mathbf{C}^{n+1}$ lying over the point $x = (x_0:\dots:x_n)$ in $\mathbf{P}_n$. Then the map $\mu: \mathbf{P}_n \to \mathbf{u}(n+1)^*$ defined by

$$\mu(x) \cdot a = (2\pi i\|x^*\|^2)^{-1}\ \bar{x}^{*t}\, a x^* \qquad \text{for } a \in \mathbf{u}(n+1)$$

is a moment map for the action of $U(n+1)$ on $\mathbf{P}_n$. Moreover μ is uniquely determined up to the addition of a scalar multiple of the trace.

Proof. The Lie algebra of $\mathbf{u}(n+1)$ decomposes as

$$\mathbf{u}(n+1) = \mathbf{su}(n+1) \oplus i\mathbf{R}1_{n+1}$$

where 1_{n+1} is the identity matrix. Here $\mathbf{su}(n+1)$ is the Lie algebra of the special unitary group, which is semisimple, while $i\mathbf{R}1_{n+1}$ is the Lie algebra of the central one-parameter compact subgroup of $U(n+1)$ which acts trivially on $\mathbf{P}_n$. The projection of $\mathbf{u}(n+1)$ onto $i\mathbf{R}1_{n+1}$ is given by $a \to \mathrm{tr}(a)(n+1)^{-1} 1_{n+1}$. Thus any moment map for $SU(n+1)$ is unique, and a moment map for $U(n+1)$ is unique up to the addition of a scalar multiple of the trace.

Clearly the formula given for μ is independent of the choice of x^* and satisfies

$$\begin{aligned} \mu(kx) \cdot a &= (2\pi i\|x^*\|^2)^{-1}\, \overline{x}^{*t}\, \overline{k}^{t}\, ak\, x^* \\ &= \mu(x) \cdot k^{-1} ak = \mathrm{Ad}^*k\, \mu(x) \cdot a \end{aligned}$$

for all $k \in \mathbf{u}(n+1)$ and $a \in \mathbf{u}(n+1)$. So μ is $U(n+1)$-equivariant.

In particular since $U(n+1)$ acts transitively on P_n, to prove that 2.3 holds it suffices to consider the point $p = (1:0:\ldots:0)$. The Kähler form at this point is given by

$$\omega_p = i/2\pi \sum_{j=1}^{n} dx_j \wedge d\overline{x}_j$$

with respect to local coordinates $(x_1,\ldots,x_n) \to (1:x_1:\ldots:x_n)$ near p. But in these cooordinates the vector field induced by a on P_n takes the values

$(a_{10}, a_{20}, \ldots, a_{n0})$ at p. Also

$$d((2\pi i\|x^*\|^2)^{-1} \overline{x}^{*t} ax^*) = (2\pi i)^{-1} \sum_{j=1}^{n} (a_{0j} dx_j + a_{j0} d\overline{x}_j)$$

$$= \quad i/2\pi \sum_{j=1}^{n} (\overline{a}_{j0} dx_j - a_{j0} d\overline{x}_j)$$

at p. This corresponds to the vector $(a_{10}, \ldots, a_{n0})$ under the duality defined by ω at p. Hence 2.3 holds and μ is a moment map.

2.6. Remark. An alternative proof runs as follows. It is known that there is a natural homogeneous symplectic structure on any orbit in $u(n+1)^*$ of the co-adjoint action of $U(n+1)$, and that the corresponding moment map is the inclusion of the orbit in $u(n+1)^*$. (This is true for any compact group K by [Ar] p. 322). We can identify $\mathbf{u}(n+1)^*$ with $\mathbf{u}(n+1)$ using the standard invariant inner product on $u(n+1)$. Then the map from P_n to $u(n+1)$ given by

$$x \to (2\pi i\|x^*\|^2)^{-1} x^* \overline{x}^{*t}$$

is a $U(n+1)$-invariant symplectic isomorphism from P_n to the orbit of the skew-hermitian matrix $(2\pi i)^{-1}$ diag $(1,0,\ldots,0)$. For $x^* \overline{x}^{*t}$ is hermitian of rank 1 with x^* as an eigenvector with eigenvalue $|x^*|^2$. Lemma 2.5 follows from this because the inner product of $x^* \overline{x}^{*t}$ with any $a \in \mathbf{u}(n+1)$

is $\bar{x}^{*t} a x^*$.

To sum up: by 2.4 and 2.5, given a nonsingular complex projective variety $X \subseteq P_n$ and a compact group K acting on X by a homomorphism $\phi: K \to U(n+1)$, a moment map $\mu: X \to \mathbf{k}^*$ is defined by

2.7. $$\mu(x) \cdot a = (2\pi i \|x^*\|^2)^{-1} \bar{x}^{*t} \phi_*(a) x^*$$

for each $a \in \mathbf{k}$ and $x \in X$. This moment map is functorial in X and K.

2.8. Consider the example 2.2 of configurations of points on the complex sphere $\mathbf{P}_1$ acted on by $SU(2)$. The Lie algebra of $SU(2)$ is isomorphic to $\mathbf{R}^3$, and $\mathbf{P}_1$ can be identified with the unit sphere in $\mathbf{R}^3$ in such a way that the moment map $\mu: (\mathbf{P}_1)^n \to \mathbf{su}(2)$ sends a configuration of n points on the sphere to its centre of gravity in $\mathbf{R}^3$ (up to a scalar factor of n).

Henceforth we shall assume that <u>a moment map</u> μ <u>exists for the action of</u> K <u>on</u> X.

Fix an inner product on the Lie algebra $\mathbf{k}$ which is invariant under the adjoint action of K and denote the product of a and b by $a.b$. Use this product to identify $\mathbf{k}$ with its dual $\mathbf{k}^*$.

For example if $K \subseteq U(n+1)$ we can take the restriction to $\mathbf{k}$ of the standard inner product given by

$$a.b = -\mathrm{tr}(ab)$$

on $\mathbf{u}(n+1)$. Then 2.6 implies that for each $x \in X$ the element $\mu(x)$ of $\mathbf{k}^*$

is identified with the orthogonal projection of the skew-hermitian matrix $(2\pi i\|x^*\|^2)^{-1} x^* \bar{x}^{*t}$ onto $\mathbf{k}$.

Also choose a K-invariant Riemannian metric on X. If X is Kähler (in particular if X is a projective variety) then the natural choice is the real part of the Kähler metric on X.

2.9. Definition. Let $f\colon X \to \mathbf{R}$ be the function given by

$$f(x) = \|\mu(x)\|^2$$

for $x \in X$, where $\| \ \|$ is the norm on $\mathbf{k}$ induced by the fixed inner product.

We want to consider the function $f\colon X \to \mathbf{R}$ as a Morse function on X. For any $x \in X$ let $\{x_t \mid t \geq 0\}$ be the trajectory of $-\mathrm{grad}\, f$ such that $x_0 = x$, i.e. the path of steepest descent of f starting from x. Let

$$\omega(x) = \{y \in X \mid \text{every neighbourhood of } y \text{ in } X \text{ contains points } x_t \text{ for } t \text{ arbitrarily large}\}$$

be the set of limit points of the trajectory as $t \to \infty$. Then $\omega(x)$ is closed and nonempty (since X is compact) and is connected. For suppose there are disjoint open sets U and V in X such that $\omega(x) \subseteq U \cup V$. Then for every $y \notin U \cup V$ there is some $t_y \geq 0$ and a neighbourhood W_y of y such that $x_t \notin W_y$ for $t \geq t_y$. But $X - (U \cup V)$ is compact so there is some $T > 0$ such that $t \geq T$ implies $x_t \in U \cup V$. Since the set $\{x_t \mid t \geq T\}$ is connected it is contained either in U or in V, and thus $\omega(x)$ is also contained in either U or V. We conclude that

2.10. For every $x \in X$ the limit set $\omega(x)$ is connected. Also every point of $\omega(x)$ is critical for f.

If f were a nondegenerate Morse function in the sense of Bott (see [A & B] §1) then the set of critical points for f on X would be a finite disjoint union of connected submanifolds $\{C \in \mathbf{C}\}$ of X. Given such a function, 2.10 implies that for every $x \in X$ there is a unique $C \in \mathbf{C}$ such that $\omega(x)$ is contained in C. The Morse stratum S_C corresponding to any $C \in \mathbf{C}$ is then defined to consist of those $x \in X$ with $\omega(x)$ contained in C. The strata S_C retract onto the corresponding critical submanifolds C and form a smooth stratification of X in the following sense.

2.11. Definition. A finite collection $\{S_\beta \mid \beta \in \mathbf{B}\}$ of subsets of X form a stratification[1] of X if X is the disjoint union of the strata $\{S_\beta \mid \beta \in \mathbf{B}\}$, and there is a strict partial order $>$ on the indexing set $\mathbf{B}$ such that

$$\overline{S}_\beta \subseteq \bigcup_{\gamma \geq \beta} S_\gamma$$

for every $\beta \in \mathbf{B}$. (For the Morse stratification associated to a nondegenerate Morse function the partial order is given by

$$C > C' \text{ if } f(C) > f(C')$$

where for $C \in \mathbf{C}$, $f(C)$ is the value taken by f on C).

The stratification is smooth if every stratum S_β is a locally-closed submanifold of X (possibly disconnected).

In fact the set of critical points for the function $f = \|\mu\|^2$ has singularities in general so that f cannot be a nondegenerate Morse function in the sense of Bott. Nevertheless we shall see that the critical set of f is a finite disjoint union of closed subsets $\{C_\beta \mid \beta \in \mathbf{B}\}$ on each of which f takes a constant value. Because of 2.10 it follows that for every $x \in X$ there is a unique $\beta \in \mathbf{B}$ such that $\omega(x)$ is contained in C_β. So X is the disjoint union of subsets $\{S_\beta \mid \beta \in \mathbf{B}\}$ where $x \in X$ lies in S_β if the limit set $\omega(x)$ of its path of steepest descent for f is contained in C_β. We shall find that for a suitable Riemannian metric the subsets $\{S_\beta \mid \beta \in \mathbf{B}\}$ form a smooth K-invariant stratification of X.

2.12. Example. The norm-square of the moment map μ associated to the action of SU(2) on sequences of n points in $\mathbf{P}_1$ identified with the unit sphere in $\mathbf{R}^3$ is given by

$$(x_1,\ldots,x_n) \to \|x_1 + x_2 + \ldots + x_n\|^2,$$

where $\| \ \|$ is the usual norm on $\mathbf{R}^3$. As is always the case $\|\mu\|^2$ takes its minimum value on $\mu^{-1}(0)$ which consists of all sequences with centre of gravity at the origin. Note that if n is even $\mu^{-1}(0)$ is singular near configurations containing two sets of n/2 coincident points. One can check that the critical configurations not contained in $\mu^{-1}(0)$ are those in which some number $r > n/2$ of the n points coincide somewhere on the sphere and the other $n-r$ coincide at the antipodal point. The connected components of the set of non-minimal critical points are thus submanifolds and are indexed by subsets of $\{1,\ldots,n\}$ of cardinality greater than $n/2$. The union of the Morse strata corresponding to subsets of fixed cardinality r consists of all sequences such that precisely r of the points coincide somewhere on $\mathbf{P}_1$.

Given any smooth stratification $\{S_\beta \mid \beta \in \mathbf{B}\}$ of the manifold X one can build up the cohomology of X inductively from the cohomology of the strata. This is done by using the Thom-Gysin sequences which for each $\beta \in \mathbf{B}$ relate the cohomology groups of the stratum S_β and of the two open subsets

$$\bigcup_{\gamma<\beta} S_\gamma \quad \text{and} \quad \bigcup_{\gamma\leq\beta} S_\gamma$$

of X. These give us the famous Morse inequalities, which can be expressed as follows. For any space Y let $P_t(Y)$ be the Poincaré series given by

$$P_t(Y) = \sum_{i\geq 0} t^i \dim H^i(Y;\mathbf{Q}).$$

Assume for convenience that if $\beta \in \mathbf{B}$ then each component of the stratum

S_β has the same codimension, $d(\beta)$ say, in X. Then the Morse inequalities say that

2.13.

$$\sum_{\beta \in \mathbf{B}} t^{d(\beta)} P_t(S_\beta) - P_t(X) = (1+t)R(t)$$

where $R(t)$ is a series with non-negative integer coefficients (see [A & B] §1).

2.14. A smooth stratification of X is called <u>perfect</u> if the Morse inequalities are equalities, that is if

$$P_t(X) = \sum_{\beta \in \mathbf{B}} t^{d(\beta)} P_t(S_\beta) \ .$$

When the stratification is induced by a nondegenerate Morse function f, one can replace $P_t(S_C)$ by $P_t(C)$ for each critical submanifold C, because the stratum S_C retracts onto C: this is the form in which the Morse inequalities are usually seen. In this form the metric does not appear in the inequalities.

If a space Y is acted on by a topological group G then the <u>equivariant cohomology</u> $H_G^*(Y;Q)$ <u>of</u> Y <u>with coefficients in</u> Q is defined by

2.15 $$H_G^*(Y;Q) = H^*(EG \times_G Y;Q)$$

where $EG \to BG$ is the universal classifying bundle for the group G, and $EG \times_G Y$ is the quotient of $EG \times Y$ by the diagonal action of G acting on EG on the right and on Y on the left.

For any smooth stratification $\{S_\beta \mid \beta \in \mathbf{B}\}$ of X whose strata S_β are all invariant under the action of the group K on X we obtain equivariant Morse inequalities

2.16

$$\sum_{\beta \in \mathbf{B}} t^{d(\beta)} P_t^K(S_\beta) - P_t^K(X) = (1+t)R(t)$$

where $R(t)$ has non-negative coefficients and P_t^K denotes the equivariant Poincare series (see [A & B] §1).

The stratification is called equivariantly perfect if these are equalities.

It will be shown that the function $f = |\mu|^2$ on X is equivariantly perfect in the sense that

2.17

$$P_t^K(X) = \sum_{\beta \in \mathbf{B}} t^{\lambda(\beta)} P_t^K(C_\beta)$$

where the sum is over the critical subsets $\{C_\beta \mid \beta \in \mathbf{B}\}$ and $\lambda(\beta)$ is the index of f along C_β. This is done by showing that if X is given a suitable metric then the stratification $\{S_\beta \mid \beta \in \mathbf{B}\}$ induced by f is equivariantly perfect and each stratum S_β retracts equivariantly onto the corresponding critical subset C_β.

We shall finish this section with a criterion due to Atiyah and Bott for a stratification to be equivariantly perfect (see [A & B] 1.4 and the corollary following 1.8).

2.18. Lemma. Suppose $\{S_\beta \mid \beta \in \mathbf{B}\}$ is a smooth K-invariant stratification of X such that for each $\beta \in \mathbf{B}$ the equivariant Euler class of the normal bundle to S_β in X is not a zero-divisor in $H_K^*(S_\beta;Q)$. Then the stratification is equivariantly perfect over $\mathbf{Q}$.

Proof. We need to show that the equivariant Thom-Gysin sequences

$$\cdots \to H_K^{n-d(\beta)}(S_\beta;Q) \to H_K^n(\bigcup_{\gamma \leq \beta} S_\gamma;Q) \to H_K^n(\bigcup_{\gamma < \beta} S_\gamma;Q) \to \cdots$$

split into short exact sequences for all $\beta \in \mathbf{B}$. It is enough to show that each map

$$H_K^{n-d(\beta)}(S_\beta;Q) \to H_K^n(\bigcup_{\gamma \leq \beta} S_\gamma;Q)$$

is injective. This will certainly happen if the composition with the restriction map

$$H_K^n(\bigcup_{\gamma \leq \beta} S_\gamma;Q) \to H_K^n(S_\beta;Q)$$

is injective. But this composition is multiplication by the equivariant Euler class of the normal bundle to S_β in X. The result follows.

Remark. For ordinary cohomology this criterion can never be satisfied since the cohomology vanishes in dimensions greater than $\dim X$, and hence every cohomology class of dimension greater than zero is nilpotent.

Footnote

[1]Note that this definition does not require the decomposition of X into strata to be topologically locally trivial.

§3. Critical points for the square of the moment map

Suppose that K is a compact Lie group acting on a compact symplectic manifold X and that $\mu\colon X \to \mathbf{k}^*$ is a moment map for this action. Our aim is to use the function $f = \|\mu\|^2 : X \to \mathbf{R}$ as a Morse function on X, where $\| \ \|$ is the norm associated to any inner product on $\mathbf{k}$ which is invariant under the adjoint action of K. In this section we shall investigate the set of critical points for the function f on X.

As before if $a \in \mathbf{k}$ let $x \to a_x$ be the vector field on X induced by a.

3.1. Lemma. A point $x \in X$ is critical for f if and only if $\mu(x)_x = 0$, where $\mu(x) \in \mathbf{k}^*$ is identified with an element of $\mathbf{k}$ by using the fixed invariant inner product on $\mathbf{k}$.

Proof. Let $\{a_i \mid 1 \leq i \leq \dim\mathbf{k}\}$ be an orthonormal basis of $\mathbf{k}$ and for $1 \leq i \leq \dim \mathbf{k}$ let $\mu_i\colon X \to \mathbf{R}$ be given by $\mu_i(x) = \mu(x).a_i$. Then

$$\mu(x) = \sum_i \mu_i(x)\, a_i$$

when $\mathbf{k}^*$ is identified with $\mathbf{k}$, and

$$f(x) = \|\mu(x)\|^2 = \sum_i (\mu_i(x))^2,$$

so that $df(x) = \sum_i 2\mu_i(x)d\mu_i(x)$.

Now $df(x) \in T_x^*X$ vanishes if and only if its ω-dual in T_xX does, where ω is the symplectic form on X. But by definition 2.3 of a moment map the ω-dual of each $d\mu_i(x)$ is just the vector $(a_i)_x$. Hence the ω-dual of $df(x)$ is

3.2

$$2(\sum_i \mu_i(x)a_i)_x = 2(\mu(x))_x .$$

The result follows.

3.3. Now let T be a maximal torus of K and let $\mathbf{t}$ be its Lie algebra. Then it is easy to check that the composition μ_T: $X \to \mathbf{k}^* \to \mathbf{t}^*$ of μ with the restriction map $\mathbf{k}^* \to \mathbf{t}^*$ is a moment map for the action of T on X. When the inner product on $\mathbf{k}$ is used to identify $\mathbf{k}^*$ with $\mathbf{k}$ and $\mathbf{t}^*$ with $\mathbf{t}$ then μ_T becomes the orthogonal projection of μ onto $\mathbf{t}$. Thus if $\mu(x) \in \mathbf{t}$ then $\mu_T(x) = \mu(x)$ and hence x is critical for the function $f = \|\mu\|^2$ if and only if it is critical for the function $f_T = \|\mu_T\|^2$ by 3.1.

Therefore we shall next investigate the critical points of $f_T = \|\mu_T\|^2$.

The moment maps $\mu_T : X \to \mathbf{t}^*$ associated to torus actions on X have been studied by Atiyah in [A2]. Theorem 1 of [A2] tells us that

3.4. <u>the image under</u> μ_T <u>of the fixed point set of</u> T <u>on</u> X <u>is a finite set</u>

A <u>of points in</u> $\mathbf{t}^*$ <u>and</u> $\mu_T(X)$ <u>is the convex hull</u> Conv(**A**) <u>of</u> **A** <u>in</u> $\mathbf{t}^*$.

The elements of **A** will be called the <u>weights</u> of the symplectic action of T on X. This terminology is explained by the following example.

3.5. <u>Example.</u> Let $X \subseteq \mathbf{P}_n$ be a nonsingular complex projective variety, and let T act on X via a homomorphism $\phi: T \to U(n+1)$. By conjugating ϕ by an element of $U(n+1)$ we may assume that

$$\phi(t) = \text{diag}\,(\alpha_0(t),\ldots,\alpha_n(t)) \quad \text{for} \quad t \in T,$$

where

$$\alpha_j : T \to S^1 = \{z \in \mathbf{C}^* \mid |z| = 1\}, \qquad 0 \le j \le n$$

are characters of T whose derivatives at 1 are the weights of the representation of T on $\mathbf{C}^{n+1}$. If the tangent space at 1 to S^1 is identified with the line $2\pi i\mathbf{R}$ in $\mathbf{C}$ and hence with $\mathbf{R}$ in the usual way then the derivative of each α_j at 1 can be identified with an element of $\mathbf{t}^*$. By abuse of notation this element of $\mathbf{t}^*$ will also be denoted by α_j. Then the derivative ϕ_* of ϕ at 1 is given by

$$\phi_*(\xi) = 2\pi i\, \text{diag}\,(\xi.\alpha_0,\ldots,\xi.\alpha_n).$$

By 2.7 a moment map μ_T for T is given by

$$\mu_T(x).\xi = (2\pi i\|x^*\|^2)^{-1}\,\overline{x}^{*t}\phi_*(\xi)x^* = \|x^*\|^{-2} \sum_{0 \le j \le n} |x_j|^2\,\alpha_j.\xi$$

for each $\xi \in \mathbf{t}$, where $x^* = (x_0,\ldots,x_n)$ is any element of $\mathbf{C}^{n+1} - \{0\}$ representing x. Thus

3.6. $$\mu_T(x) = \|x^*\|^{-2} \sum_{0 \leq j \leq n} |x_j|^2 \alpha_j .$$

The point $x \in X$ is fixed by T if and only if there is some $\alpha \in t^*$ such that $\alpha_j = \alpha$ whenever $x_j \neq 0$; and then clearly $\mu_T(x) = \alpha$. So at least when X is the whole projective space $\mathbf{P}_n$ the set $\mathbf{A}$ is just the set $\{\alpha_0, \ldots, \alpha_n\}$ of weights of the representation of T on $\mathbf{C}^{n+1}$, and the formula 3.6 shows immediately that $\mu_T(\mathbf{P}_n)$ is the convex hull of $\mathbf{A}$.

We need some definitions.

3.7. Definition. For any $\beta \in \mathbf{t}$ let T_β be the closure in T of the real one-parameter subgroup $\exp \mathbf{R}\beta$. Thus T_β is a subtorus of T. Let $\mu_\beta : X \to \mathbf{R}$ be given by $\mu_\beta(x) = \mu(x).\beta$. Then by the definition of a moment map the cotangent field $x \to d\mu_\beta(x)$ on X is ω-dual to the vector field $x \to \beta_x$ induced by β on X. If $x \in X$ then $\beta_x = 0$ if and only if x is fixed by the subgroup $\exp \mathbf{R}\beta$ of T and hence by its closure T_β in T. Therefore the critical set of the function μ_β on X is precisely the fixed point set of the subtorus T_β of T. It is well known that

3.8. every connected component of the fixed point set of a torus action on X is a submanifold of X, and the induced action of the torus on its normal tangent bundle in X has no nonzero fixed vectors.

(To see this one puts a T-invariant Riemannian metric on X and uses normal coordinates).

Using this fact Atiyah shows that

3.9. μ_β <u>is a nondegenerate Morse function on</u> X <u>in the sense of Bott</u> (see lemma 2.2 of [A2]).

3.10. <u>Definition.</u> Let Z_β be the union of those connected components of the critical set of μ_β on which μ_β takes the value $\|\beta\|^2$. Thus if $x \in Z_\beta$ then $\mu(x)$ lies in the affine hyperplane in **k** containing β and perpendicular to the line from β to the origin.

Z_β is a submanifold of X (possibly disconnected) fixed by T_β and invariant under T. In fact it is a <u>symplectic submanifold</u> of X (see [G & S] lemma 3.6 or lemma 4.8 below).

3.11. <u>Example.</u> If $X \subseteq \mathbf{P}_n$ is a smooth projective variety and T acts on X via a homomorphism ϕ: $T \to U(n+1)$ then Z_β is the intersection with X of a linear subvariety of $\mathbf{P}_n$. If $\phi(t) = \text{diag}\,(\alpha_0(t),\ldots,\alpha_n(t))$ for $t \in T$, where $\alpha_0,\ldots,\alpha_n$ are characters of T identified with points of $\mathbf{t}^*$, then

$$Z_\beta = \{(x_0:\ldots:x_n) \in X \mid x_j = 0 \text{ unless } \alpha_j.\beta = \|\beta\|^2\}.$$

Note that the inner product on **t** gives **t** the structure of a normed space. For any nonempty closed convex set $C \subseteq \mathbf{t}$ there is then a unique point of minimal norm in C. This point will be called _the point of_ C _closest to the origin_ 0.

The point of these definitions is the following result.

3.12. _Lemma. Let_ x _be a point of_ X _and let_ $\beta = \mu_T(x) \in \mathbf{t}$ _where_ **t** _and its dual are identified via the inner product. Then_ x _is critical for the function_ $f_T = |\mu_T|^2$ _if and only if_ $x \in Z_\beta$; _and if this is the case then_ β _is the closest point to_ 0 _of the convex hull of some nonempty subset of the set_ **A** _of weights defined in_ 3.4.

Proof. By 3.1 x is critical if and only if $\beta_x = 0$, i.e. if and only if x is fixed by T_β. Since $\mu_\beta(x) = \mu_T(x).\beta = \|\beta\|^2$ it follows that x is fixed by T_β if and only if it lies in Z_β.

For any $z \in Z_\beta$ we have $\mu_T(z).\beta = \|\beta\|^2$ so that $\|\mu_T(z)\|^2 \geq \|\beta\|^2$ with equality if and only if $\mu_T(z) = \beta$. So β is the closest point to 0 of $\mu_T(Z_\beta)$ if $\beta \in \mu_T(Z_\beta)$ and hence if $x \in Z_\beta$. But we can apply 3.4 to the action of T on Z_β to deduce that $\mu_T(Z_\beta)$ is the convex hull of the image under μ_T of the fixed point set of T on Z_β, which is a subset of **A**. The result follows.

This lemma can be used to describe the critical set of the function $f = \|\mu\|^2$ associated to the action of the whole group K.

3.13. Definition. Let β be a point in the Lie algebra $\mathbf{t}$ of the maximal torus T of K. Then β will be called a minimal combination of weights of the action of T on X if it is the closest point to the origin of the convex hull in $\mathbf{t}$ of some nonempty subset of the set of weights $\mathbf{A}$ defined at 3.4. Let $\mathbf{t}_+$ be a fixed positive Weyl chamber in $\mathbf{t}$ and denote by $\mathbf{B}$ the set of all minimal combinations of weights which lie in $\mathbf{t}_+$.

$\mathbf{B}$ will be the indexing set for the stratification of X which we shall associate to the function f.

3.14. Definition. For $\beta \in B$ let $C_\beta = K(Z_\beta \cap \mu^{-1}(\beta))$.

Then we have

3.15. Lemma. The critical set of f on X is the disjoint union of the closed subsets $\{C_\beta \mid \beta \in \mathbf{B}\}$ of X.

Proof. As usual, identify $\mathbf{k}^*$ with $\mathbf{k}$ using the inner product.

For any $x \in X$ there is some $k \in K$ such that $\mathrm{Ad}k\mu(x) \in \mathbf{t}_+$. By the definition of a moment map $\mathrm{Ad}k\mu(x) = \mu(kx)$. Since f is a K-invariant map x is critical for f if and only if kx is for any such k. But $\mu(kx) \in \mathbf{t}$ so by 3.3 kx is critical for $f = \|\mu\|^2$ if and only if it is critical for $f_T = |\mu_T|^2$. Let $\beta = \mu(kx) \in \mathbf{t}_+$. By 3.12 kx is critical for f_T if and only if $kx \in Z_\beta$, and if this happens then $\beta \in \mathbf{B}$.

Therefore the critical set for f is the union of the closed sets $C_\beta = K(Z_\beta \cap \mu^{-1}(\beta))$ as β runs over **B**. Moreover for each $\beta \in \mathbf{B}$ the image of C_β under μ is precisely the orbit of β under the adjoint representations of K. Since any adjoint orbit in **k** intersects the positive Weyl chamber in a unique point the subsets $\{C_\beta \mid \beta \in \mathbf{B}\}$ must be disjoint. The result follows.

The subsets $\{C_\beta \mid \beta \in \mathbf{B}\}$ will therefore be called the <u>critical subsets</u> for f.

3.16. <u>Corollary.</u> <u>The image under</u> μ <u>of each connected component of the critical set for</u> f <u>is a single adjoint orbit in</u> $\mathbf{k}^* \cong \mathbf{k}$. <u>For each</u> $\beta \in \mathbf{B}$, C_β <u>consists of those critical points for</u> f <u>whose image under</u> μ <u>lies in the adjoint orbit of</u> β. <u>Thus the function</u> $f = |\mu|^2$ <u>takes the value</u> $\|\beta\|^2$ <u>on</u> C_β.

3.17. <u>Example.</u> If $X = (\mathbf{P}_1)^n$ is acted on by SU(2) as in 2.2 then T_β is the maximal torus T of SU(2) when $\beta \neq 0$. The fixed point set of T consists of all configurations such that every point is either at 0 or ∞. Identify **t** with **R** and give it the standard inner product, so that the identity character of $T = S^1$ becomes the point 1 in **R**. Take $\mathbf{R}^+$ as the positive Weyl chamber. Then the moment map sends a configuration with r points at 0 and the rest at ∞ to $2r-n \in \mathbf{t}$. So

$$B = \{2r-n \mid \frac{1}{2} n < r \leq n\} \cup \{0\},$$

and if $\beta = 2r-n$ then Z_β consists of configurations with r points at 0 and the rest at ∞. Thus the last lemma agrees with 2.12.

§4. The square of the moment map as a Morse function

As in §2 and §3 let X be a compact symplectic manifold acted on by a compact Lie group K, and assume that a moment map $\mu: X \to \mathbf{k}^*$ exists for this action. We want to apply Morse theory to the function $f = \|\mu\|^2: X \to \mathbf{R}$, where $\| \ \|$ is the norm associated to any inner product on the Lie algebra $\mathbf{k}$ which is invariant under the adjoint action of K. Problems arise because the critical set of f has singularities. However we shall see in this section that f is a minimally degenerate function on X. It is shown in the appendix §10 that such functions are sufficiently well behaved to have associated Morse inequalities.

To show that f is minimally degenerate we need to find a minimising manifold along each of the critical subsets $\{C_\beta \mid \beta \in B\}$ defined in §3. That is, for each $\beta \in \mathbf{B}$ we require a submanifold Σ_β of some neighbourhood of C_β with orientable normal bundle in X and such that the restriction of f to Σ_β takes its minimum value on C_β. We also require that for each $x \in C_\beta$ the tangent space $T_x\Sigma_\beta$ is maximal among subspaces of T_xX on which the Hessian $H_x(f)$ is positive semi-definite.

First fix a K-invariant Riemannian metric on X.

4.1. Note that such a metric and the symplectic structure give X a K-invariant almost-complex structure as follows. The metric can be used to identify the symplectic form with a skew-adjoint linear operator A on the tangent bundle TX. Then $A^2 = -AA^*$; and since AA^* is self-adjoint with positive eigenvalues it has a unique positive definite square root $(AA^*)^{1/2}$. If we rescale the metric by $(AA^*)^{-1/2}$ then A is replaced by $J = A(AA^*)^{-1/2}$ which satisfies $J^2 = -1$. Hence there is a complex structure on TX such that J is multiplication by i.

We can thus assume that the chosen K-invariant metric on X has been suitably normalised so that

4.2. X <u>has a K-invariant almost-complex structure such that if</u> $\xi \in T_xX$ <u>then</u> $i\xi$ <u>is the dual with respect to the metric of the linear form</u> $\zeta \mapsto \omega_x(\zeta,\xi)$ <u>on</u> T_xX.

Note that this implies that

4.3. $\operatorname{grad} \mu_\beta(x) = i\beta_x$ for every $x \in X$,

because by the definition of a moment map the cotangent vector field $d\mu_\beta$ on X is ω-dual to the tangent vector field $x \mapsto \beta_x$.

4.4 <u>Remark.</u> When X is a Kåhler manifold the real part of the Kåhler metric is the obvious choice for a Riemannian metric on X. The induced almost-complex structure then coincides with the complex structure of X. In this section where X is merely a symplectic manifold, the almost-complex structure is used not only for convenience but also because it links up with the work of later sections on Kåhler manifolds.

Recall from lemma 3.15 that the set of critical points for f on X is the disjoint union of the closed subsets $\{C_\beta \mid \beta \in \mathbf{B}\}$. The indexing set $\mathbf{B}$ is the set of minimal weight combinations in the positive Weyl chamber as defined at 3.13. For each β in $\mathbf{B}$ the critical subset C_β is $K(Z_\beta \cap \mu^{-1}(\beta))$ where Z_β is the symplectic submanifold of X defined at 3.10. This submanifold Z_β is the union of certain components of the fixed point set of the subtorus T_β generated by β, or equivalently of the critical set of the function μ_β on X (which is a nondegenerate Morse function in the sense of Bott). The components contained in Z_β are those on which μ_β takes the value $\|\beta\|^2$.

4.6. For each $\beta \in \mathbf{B}$ there is a Morse stratum Y_β associated to Z_β which consists of all points of X whose paths of steepest descent under μ_β have limit points in Z_β. This Morse stratum Y_β (which depends on the chosen metric) is a locally-closed submanifold of X. (These facts are well known; a proof is given in the appendix, but this covers the more general case of minimally degenerate functions, which are harder to deal with than nondegenerate ones such as μ_β).

4.7. Example. Consider again the projective variety $X = (\mathbf{P}_1)^n$ acted on diagonally by SU(2). We have seen at 3.17 that the nonzero elements of the indexing set $\mathbf{B}$ may be identified with integers r such that $n/2 < r \leq n$, and that Z_r consists of sequences of points in $\mathbf{P}_1$ of which r lie at 0 and the rest lie at ∞. It is not hard to see that Y_r consists of all sequences of points precisely r of which lie at 0.

Note that KY_r thus consists of all sequences of points such that r points and no more coincide somewhere on $\mathbf{P}_1$. By 2.12 this is exactly the Morse stratum indexed by r for the function $\|\mu\|^2$ on X.

Recall that we need a minimising manifold Σ_β along each critical subset C_β. It will be shown that we can take Σ_β to be an open subset of KY_β.

<u>Remark.</u> It will then follow from theorem 10.4 that the Morse stratum S_β coincides with KY_β in a neighbourhood of C_β. In fact in the Kåhler case we shall see that

$$S_\beta = K\, Y_\beta^{min}$$

where Y_β^{min} is a certain open subset of Y_β. If one were only interested in the Kåhler case it would be possible to avoid minimising manifolds and simplify the appendix somewhat by using this fact. When X is just a symplectic manifold the equality above does not hold for every invariant metric on X. For example, consider $X = (\mathbf{P}_1)^n$ with symplectic form $\omega\oplus\ldots\oplus\omega$ and metric $2\rho\oplus\rho\oplus\ldots\oplus\rho$ where ω and ρ are the usual symplectic form and metric on $\mathbf{P}_1$. It may always be possible to choose a metric for which the equality holds, or at least always when $\pi_1(X) = 0$ (that would follow immediately if it were shown that every simply-connected compact symplectic manifold is Kåhler[1]) but I cannot prove it.

First, in order to show that KY_β is smooth near C_β, we must investigate what elements of K preserve Y_β.

4.8. Definition. For each $\beta \in \mathbf{B}$ let $\mathrm{Stab}\beta = \{k \in K \mid \mathrm{Ad}\, k(\beta) = \beta\}$ be the stabiliser of β in K. $\mathrm{Stab}\beta$ is also the centraliser of the subtorus T_β in K, so that it is connected if K is connected (by [He] VII Cor. 2.8) and is a compact subgroup of K. Let $\mathrm{stab}\beta = \{a \in \mathbf{k} \mid [a,\beta] = 0\}$ be the Lie algebra of $\mathrm{Stab}\beta$.

$\mathrm{Stab}\beta$ acts on the symplectic submanifold Z_β of X, and the composition of μ restricted to Z_β with the orthogonal projection of $\mathbf{k}$ onto $\mathrm{stab}\beta$ is a moment map for this action (cf. 3.3); as usual $\mathbf{k}$ and its dual are identified via the inner product. But if $x \in Z_\beta$ then T_β fixes x and hence also fixes $\mu(x)$ since μ is a K-equivariant map. Therefore $\mu(x) \in \mathrm{Stab}\beta$. It follows that

4.9. The restriction of μ to Z_β maps Z_β to $\mathrm{stab}\beta$ and can be regarded as a moment map for the action of $\mathrm{Stab}\beta$ on Z_β.

In order to show that KY_β is smooth in a neighborhood of $C_\beta = K(Z_\beta \cap \mu^{-1}(\beta))$ we need the following lemma.

4.10. Lemma. If $x \in Z_\beta \cap \mu^{-1}(\beta)$ then

$$\{k \in K \mid kx \in Y_\beta\} = \mathrm{Stab}\beta$$

and

$$\{a \in \mathbf{k} \mid a_x \in T_x Y_\beta\} = \mathrm{stab}\beta.$$

<u>Proof.</u> It is clear from the definitions that Z_β is invariant under $\mathrm{Stab}\beta$ and that $Z_\beta \subseteq Y_\beta$. It follows that

$$\mathrm{Stab}\beta \subseteq \{k \in K \mid kx \in Y_\beta\}$$

and

$$\mathrm{stab}\beta \subseteq \{a \in \mathbf{k} \mid a_x \in T_x Y_\beta\}.$$

On the other hand suppose $k \in K$ is such that $kx \in Y_\beta$. Then the path of steepest descent from kx for the function μ_β has a limit point in Z_β, and by definition μ_β takes the value $\|\beta\|^2$ on Z_β. Thus as $\mu_\beta(kx) = \mu(kx).\beta$ we have $\mu(kx).\beta \geq \|\beta\|^2$. But

$$\|\mu(kx)\|^2 = \|\mu(kx)\|^2 = \|\beta\|^2.$$

Together these imply that $\mu(kx) = \beta$, and since $\mu(kx) = \mathrm{Ad}k\mu(x) = \mathrm{Ad}k\beta$ it follows that $k \in \mathrm{Stab}\beta$.

Now suppose $a \in \mathbf{k}$ is such that $a_x \in T_x Y$. For $t \in \mathbf{R}$ we have

$$\mu((\exp ta)x) = \beta + td\mu(x)(a_x) + e(t)$$

where $e(t) = O(t^2)$ as $t \to 0$; and

$$d\mu(x)(a_x) = [a,\mu(x)] = [a,\beta]$$

because μ is K-equivariant. As $[a,\beta].\beta = a.[\beta,\beta] = 0$, it follows that

$$\mu_\beta((\exp ta)x) = \|\beta\|^2 + \beta.e(t).$$

But also $\|\mu(\exp ta)x\|^2 = \|\mu(x)\|^2 = \|\beta\|^2$ for all $t \in \mathbf{R}$, so that

$$\|\beta\|^2 = \|\beta + t[a,\beta] + e(t)\|^2$$

$$= \|\beta\|^2 + t^2\|[a,\beta]\|^2 + 2\beta.e(t) + O(t^3) \quad \text{as} \quad t \to 0.$$

Thus

$$2\beta.e(t) = -t^2\|[a,\beta]\|^2 + O(t^3) \quad \text{as} \quad t \to 0$$

and hence

$$\mu_\beta((\exp ta)x) = \|\beta\|^2 - \frac{1}{2} t^2\|[a,\beta]\|^2 + O(t^3) \quad \text{as} \quad t \to 0.$$

But by assumption $a_x \in T_x Y_\beta$ which is the sum of the nonnegative eigenspaces of the Hessian $H_x(\mu_\beta)$ of μ_β at x because μ_β is nondegenerate in the sense of Bott. The last equation shows that this is impossible unless $[a,\beta] = 0$, i.e. unless $a \in \mathrm{stab}\beta$. This completes the proof.

4.11. Corollary. *The subset KY_β of X is a smooth submanifold when restricted to some K-invariant neighbourhood of $C_\beta = K(Z_\beta \cap \mu^{-1}(\beta))$ in X.*

Proof. Since Y_β is invariant under $\mathrm{Stab}\beta$, the map $\sigma\colon K \times Y_\beta \to X$ given by $\sigma(k,x) = kx$ induces a map $\tilde{\sigma}\colon K \times_{\mathrm{Stab}\beta} Y_\beta \to X$ whose image is KY_β. It is easily checked from the definition of Y_β that if $\varepsilon > 0$ is sufficiently small the subset $\{y \in Y_\beta \mid \mu_\beta(x) \le \|\beta\|^2 + \varepsilon\}$ of Y_β is a compact neighbourhood of Z_β in Y_β. Moreover its complement in Y_β is contained in the subset $\{y \in X \mid \|\mu(y)\| \ge \|\beta\| + \|\beta\|^{-1} \varepsilon\}$ of X, which is closed, K-invariant and does not meet $Z_\beta \cap \mu^{-1}(\beta)$. From this one can deduce easily that if $x \in Z_\beta \cap \mu^{-1}(\beta)$ then $\tilde{\sigma}$ maps each neighbourhood of the point in $K \times_{\text{Stab}\beta} Y_\beta$ represented by $(1,x)$ onto a neighbourhood of x in the image KY_β of $\tilde{\sigma}$.

The derivative of σ at any point of the form $(1,x)$ sends a vector (a,ξ) in $\mathbf{k} \times T_x Y_\beta$ to the tangent vector $a_x + \xi \in T_x X$. The tangent space of $K \times_{\text{Stab}\beta} Y_\beta$ at a point represented by $(1,x)$ is the quotient of $\mathbf{k} \times T_x Y_\beta$ by the subspace consisting of all (a,ξ) such that $a \in \text{Stab}_\beta$ and $\xi = -a_x$. Thus 4.11 shows that the derivative of $\tilde{\sigma}$: $K \times_{\text{Stab}\beta} Y_\beta \to X$ is injective at a point represented by $(1,x)$ with $x \in Z_\beta \cap \mu^{-1}(\beta)$, and hence also in some neighbourhood V of this point. The preceding paragraph shows that the image $\tilde{\sigma}(V)$ of V under $\tilde{\sigma}$ is a neighbourhood of x in KY_β. Therefore it follows from the inverse function theorem that the image KY_β of $\tilde{\sigma}$ is smooth in some neighbourhood of x.

We have thus shown that KY_β is smooth near $Z_\beta \cap \mu^{-1}(\beta)$. It follows that KY_β is smooth in some K-invariant neighbourhood of $C_\beta = K(Z_\beta \cap \mu^{-1}(\beta))$, as required.

We are aiming to show that the intersection Σ_β of KY_β with a sufficiently small neighbourhood of C_β is a minimising manifold for f along C_β. The last corollary shows that the condition that Σ_β be a locally-closed submanifold of X can be satisfied. For the other conditions we need two technical lemmas.

4.12. <u>Lemma.</u> Z_β <u>is an almost-complex submanifold of</u> X. <u>Moreover</u>

T_xY_β is a complex subspace of T_xX for every $x \in Z_\beta$.

Proof. Suppose $x \in Z_\beta$. Then the compact torus T_β generated by β acts on the tangent space T_xX, which decomposes into the sum

$$V_0 \oplus V_1 \oplus \dots \oplus V_p$$

of complex subspaces where V_0 is fixed by T_β and is the tangent space to Z_β while for each $j \geq 1$ T_β acts on V_j as scalar multiplication by some nontrivial character. Thus β acts on each V_j as multiplication by some scalar $i\lambda_j$ with $\lambda_0 = 0$ and λ_j real and nonzero for $j \geq 1$. Also by 4.3 we have grad $\mu_\beta(y) = i\beta_y$ for all $y \in Y$. Therefore the Hessian $H_x(\mu_\beta)$ of μ_β at x acts on V_j as multiplication by λ_j (cf. [A2] lemma 2.2). Thus $T_xZ_\beta = V_0$ and T_xY_β is the sum of those V_j such that $\lambda_j \geq 0$, so both are complex subspaces of T_xX. The result follows.

4.13. Lemma. Suppose $x \in C_\beta = K(Z_\beta \cap \mu^{-1}(\beta))$. Then the restriction of the symplectic form ω_x to $T_x(KY_\beta)$ is nondegenerate.

Proof. First note that by 4.11 KY_β is smooth near x so $T_x(KY_\beta)$ exists. Moreover since ω is invariant under K we may assume that $x \in Z_\beta \cap \mu^{-1}(\beta)$, and then $T_x(KY_\beta) = \mathbf{k}_x + T_xY_\beta$. So any element of $T_x(KY_\beta)$ may be written in the form $a_x + \xi$ where $\xi \in T_xY_\beta$ and $a \in \mathbf{k}$ is such that a_x is orthogonal to T_xY_β (with respect to the Riemannian metric on X). Suppose that $\omega_x(a_x + \xi, \zeta) = 0$ for all $\zeta \in T_x(KY_\beta)$.

By 4.12 $i\xi \in T_xY_\beta$, so if $\langle\,,\rangle$ denotes the metric then

$$\begin{aligned} 0 &= \omega_x(a_x + \xi, i\xi) \\ &= \langle a_x + \xi, \xi \rangle \qquad \text{by 3.19,} \\ &= \langle \xi, \xi \rangle \qquad \text{by the assumption on } a. \end{aligned}$$

Hence $\xi = 0$. But then as $\mathbf{k}_x \subseteq T_x(KY_\beta)$

$$0 = \omega_x(a_x, b_x) = d\mu(x)(a_x).b$$

for every $b \in \mathbf{k}$ (see 2.3), so

$$0 = d\mu(x)(a_x) = [a, \beta]$$

because $\mu(x) = \beta$. Thus $a \in \mathrm{stab}\beta$ and hence $a_x \in T_xY_\beta$ by 4.10. But by assumption a_x is orthogonal to T_xY_β so $a_x = 0$.

This completes the proof.

4.14. Remark. Lemma 4.13 implies that there is an open neighbourhood Σ_β of the critical subset C_β in KY_β such that the restriction of the symplectic form ω to the tangent bundle $T\Sigma_\beta$ is nondegenerate. It follows that ω and the metric together induce a K-invariant almost-complex structure on Σ_β (cf. 4.1). It also follows that the normal bundle Σ_β in X can be identified with the ω-orthogonal complement $T\Sigma_\beta{}^\perp$ to $T\Sigma_\beta$ in the restriction of TX to Σ_β. Since ω is nondegenerate on $T\Sigma_\beta{}^\perp$ it gives a complex structure to this normal bundle as well.

At last we are in a position to prove

4.15. Proposition. There is a K-invariant open neighbourhood Σ_β of C_β in KY_β which is a minimising manifold for f along C_β.

Proof. It follows from 4.11 that there is a K-invariant neighbourhood Σ_β of C_β in KY_β which is smooth.

To show that Σ_β satisfies the definition of a minimising manifold, we must check that the restriction of f to Σ_β takes its minimum value on C_β. But if $x \in Y_\beta$ then its path of steepest descent under μ_β converges to a point of Z_β, and by definition Σ_β takes the value $\|\beta\|^2$ on Z_β. Hence $\mu(x).\beta = \mu_\beta(x) \geq \|\beta\|^2$ and so $f(x) = \|\mu(x)\|^2 \geq \|\beta\|^2$, which is the value taken by f on $C_\beta = K(Z_\beta \cap \mu^{-1}(\beta))$. Since f is K-invariant the same is true for $x \in KY_\beta$ and hence for $x \in \Sigma_\beta$. Moreover equality holds if and only if $x \in C_\beta$.

It follows immediately that if $x \in C_\beta$ then the restriction to $T_x\Sigma_\beta$ of the Hessian $H_x(f)$ of f at x is positive semi-definite. So it remains to show that the restriction of the Hessian $H_x(f)$ to some complementary subspace to $T_x\Sigma_\beta$ in T_xX is negative definite. As $C_\beta = K(Z_\beta \cap \mu^{-1}(\beta))$ and as the metric, Σ_β and f are all K-invariant it is enough to consider points $x \in Z_\beta \cap \mu^{-1}(\beta)$.

By lemma 4.13 the restriction of the symplectic form ω at x to $T_x\Sigma_\beta$ is nondegenerate. Therefore the orthogonal complement $T_x\Sigma_\beta^\perp$ to $T_x\Sigma_\beta$ in

T_xX with respect to ω_x is a complementary subspace to $T_x\Sigma_\beta$.

By the definition of a moment map, if $\xi \in T_xX$ and $a \in \mathbf{k}$ then

$$d\mu(x)(\xi).a = \omega_x(\xi, a_x).$$

So if ξ is ω-orthogonal to the image $\mathbf{k}_x$ in T_xX of the Lie algebra $\mathbf{k}$ then $d\mu(x)(\xi).a = 0$ for all $a \in \mathbf{k}$ and so $d\mu(x)(\xi) = 0$. Since Σ_β is invariant under K we have $\mathbf{k}_x \subseteq T_x\Sigma_\beta$, so a fortiori if $\xi \in T_x\Sigma_\beta{}^\perp$ then $d\mu(x)(\xi) = 0$.

Let $\text{Exp}\colon TX \to X$ be the exponential map associated to the metric. Then if $\xi \in T_xX$ and $t \in \mathbf{R}$,

$$\mu(\text{Exp}\, t\xi) = \beta + t d\mu(x)(\xi) + e_\xi(t)$$

where $e_\xi(t) = O(t^2)$ as $t \to 0$. Therefore if $\xi \in T_x\Sigma_\beta{}^\perp$ then

$$\mu(\text{Exp}\, t\xi) = \beta + e_\xi(t),$$

so that

$$f(\text{Exp}\, t\xi) = \|\beta + e_\xi(t)\|^2 = \|\beta\|^2 + 2\beta.e_\xi(t) + O(t^3)$$

as $t \to 0$. On the other hand

$$\mu_\beta(\text{Exp}\, t\xi) = \mu(\text{Exp}\, t\xi).\beta = \|\beta\|^2 + \beta.e_\xi(t).$$

It follows that the Hessians $H_x(f)$ and $H_x(\mu_\beta)$ agree up to a scalar factor of 2 on the subspace $T_x\Sigma_\beta{}^\perp$ of T_xX. But Σ_β is an open subset of KY_β, so $T_x\Sigma_\beta{}^\perp \subseteq T_xY_\beta{}^\perp$. Moreover by definition Y_β is the Morse stratum of the function μ_β associated to the critical submanifold Z_β, which implies that the restriction of the Hessian $H_x(\mu_\beta)$ to $T_xY_\beta{}^\perp$ is negative definite. Thus the restriction of $H_x(f)$ to $T_x\Sigma_\beta{}^\perp$ is also negative definite.

Therefore Σ_β is a minimising manifold for f along C_β as required.

We have thus shown that the function $f = \|\mu\|^2$ is minimally degenerate along each critical subset C_β. By theorem 10.2 of the appendix this implies the existence of Morse inequalities for f, and also of equivariant Morse inequalities. Indeed, by theorem 10.4 and lemma 10.5 we have the following result.

4.16. Theorem. Let X be a compact symplectic manifold acted on by a compact Lie group K, and suppose $\mu: X \to \mathbf{k}^*$ is a moment map for this action. Fix an invariant inner product on **k**. Then the set of critical points for the function $f = \|\mu\|^2$ is a finite disjoint union of closed subsets $\{C_\beta \mid \beta \in \mathbf{B}\}$, on each of which f takes a constant value. There is a smooth stratification $\{S_\beta \mid \beta \in \mathbf{B}\}$ of X such that a point $x \in X$ lies in the stratum S_β if and only if the limit set of the path of steepest descent for $f = \|\mu\|^2$ from x (with respect to a suitable K-invariant metric) is contained in C_β. For each $\beta \in \mathbf{B}$ the inclusion of C_β in S_β is an equivalence of (Čech) cohomology and also K-equivariant cohomology.

Theorem 10.4 shows in addition that

4.17. if $\beta \in \mathbf{B}$ then the stratum S_β coincides in a neighbourhood of C_β

with the minimising manifold Σ_β (which is an open subset of KY_β where Y_β is defined as at 4.6). In particular if $x \in Z_\beta \cap \mu^{-1}(\beta)$, then

$$T_x S_\beta \supseteq T_x Z_\beta .$$

From this together with remark 4.14 we deduce that

4.18. both the tangent bundle and the normal bundle to each stratum S_β have K-invariant complex structures in some neighbourhood of the critical set C_β.

Theorem 4.16 implies immediately the existence of equivariant Morse inequalities for the function $f = \|\mu\|^2$. We shall not state these explicitly until the next section, where it will be shown that they are in fact equalities.

We shall conclude this section with some remarks about the codimensions of the components of the strata S_β and the equivariant cohomology of the critical sets C_β.

Recall that when stating the Morse inequalities induced by a smooth stratification of X in §2, we made the simplifying assumption that every stratum was connected and hence had a well-defined codimension in X. In fact the stratification $\{S_\beta \mid \beta \in \mathbf{B}\}$ defined in theorem 4.16 may contain disconnected strata. Therefore it is necessary to refine it so that the components of any stratum all have the same codimension.

For $\beta \in \mathbf{B}$, the critical subset C_β was defined at 3.14 by

$$C_\beta = K(Z_\beta \cap \mu^{-1}(\beta))$$

where Z_β is the union of certain components of the critical set of the nondegenerate Morse function μ_β. Recall that the index of the Hessian $H_x(\mu_\beta)$ at any critical point x for μ_β is the dimension of any subspace of the tangent space T_xX to which the restriction of $H_x(\mu_\beta)$ is negative definite and which is maximal with this property. This is the same as the codimension of a maximal subspace of T_xX on which $H_x(\mu_\beta)$ is positive semi-definite. Since μ_β is a nondegenerate Morse function in the sense of Bott the index of $H_x(\mu_\beta)$ is constant along any component of the critical set of μ_β. Its value is called the index of μ_β along this component. So we can make the following definition.

4.19. Definition. For any integer $m \geq 0$, let $Z_{\beta,m}$ be the union of those connected components of Z_β along which the index of μ_β is m. Let $C_{\beta,m} = K(Z_{\beta,m} \cap \mu^{-1}(\beta))$.

Then each $Z_{\beta,m}$ is a symplectic submanifold of X, and C_β is the disjoint union of the closed subsets $\{C_{\beta,m} | 0 \leq m \leq \dim X\}$. (The fact that these are disjoint comes from 4.10).

The point of this definition is the following

4.20. Lemma. The index of the Hessian $H_x(f)$ of the function $f = \|\mu\|^2$

at any point $x \in C_{\beta,m}$ is

$$d(\beta,m) = m - \dim K + \dim \mathrm{Stab}\beta.$$

This is the codimension of the component which contains x of the stratum S_β.

Proof. By 4.17 the stratum S_β coincides in a neighbourhood of x with the minimising manifold Σ_β for f along C_β. It follows immediately from the definition of a minimising manifold that the index of the Hessian $H_x(f)$ equals the codimension of the component of Σ_β containing x. Thus it suffices to show that the component of Σ_β containing x has codimension $d(\beta,m)$ in X.

Since $C_{\beta,m} = K(Z_{\beta,m} \cap \mu^{-1}(\beta))$ and everything is invariant under K we may assume $x \in Z_{\beta,m} \cap \mu^{-1}(\beta)$. By definition the minimising manifold Σ_β is an open subset of KY_β where Y_β is the Morse stratum consisting of all points in X whose paths of steepest descent under the function μ_β have limit points in Z_β. Since μ_β is a nondegenerate Morse function, locally Y_β is a submanifold of X whose codimension is equal to the index of the Hessian $H_x(\mu_\beta)$ of μ_β at x. By the definition of $Z_{\beta,m}$ this index is m.

In the proof of 4.11, we saw that KY_β is locally diffeomorphic to $K \times_{\mathrm{Stab}\beta} Y_\beta$ near x. Therefore its codimension is

$$d(\beta,m) = m - \dim K + \dim \ \mathrm{Stab}\beta.$$

The result follows.

It is easy to see that for each $\beta \in \mathbf{B}$,

4.21. the critical subset $C_\beta = K(Z_\beta \cap \mu^{-1}(\beta))$ is homeomorphic to $K \times_{Stab\beta} (Z_\beta \cap \mu^{-1}(\beta))$.

(By 4.10 for each $x \in Z_\beta \cap \mu^{-1}(\beta)$ the set $\{k \in K | kx \in Z_\beta \cap \mu^{-1}(\beta)\}$ is just the subgroup $Stab\beta$ of K. Thus there is a continuous bijection.

$$K \times_{Stab\beta} (Z_\beta \cap \mu^{-1}(\beta)) \to C_\beta$$

which must be a homeomorphism since both spaces are compact and Hausdorff.)

As $Z_{\beta,m}$ is also preserved by $Stab\beta$ we deduce that

4.22. each $C_{\beta,m}$ is homeomorphic to $K \times_{Stab\beta} (Z_{\beta,m} \cap \mu^{-1}(\beta))$.

It follows immediately (see [A & B] §13) that

4.23. the K-equivariant rational cohomology $H_K^*(C_\beta;Q)$ is isomorphic to the $Stab\beta$-equivariant rational cohomology of $Z_\beta \cap \mu^{-1}(\beta)$, and similarly that

$$H_K^*(C_{\beta,m};Q) \cong H^*_{Stab\beta} (Z_{\beta,m} \cap \mu^{-1}(\beta);Q)$$

for each m. Indeed, rational coefficients are not necessary here. Any field of coefficients may be used.

We now have all the ingredients for writing down the equivariant Morse inequalities and proving that they are in fact equalities. This will be done in the next section.

Footnote

[1]Thurston has shown that there exist compact symplectic manifolds which are not Kähler, but his examples are not simply-connected.

§5. Cohomological Formulae

As in the previous sections we suppose that X is a compact symplectic manifold acted on by a compact Lie group K, that there is a fixed invariant inner product on the Lie algebra **k**, and that $\mu: X \to \mathbf{k}^* \cong \mathbf{k}$ is a moment map for the action of K on X. In the last section we saw that the function $f = \|\mu\|^2$ is a minimally degenerate Morse function on X. This implies the existence of Morse inequalities for f. In this section we shall show that these Morse inequalities calculated for rational K-equivariant cohomology are in fact equalities. Thus the function $f = \|\mu\|^2$ is equivariantly perfect for rational cohomology.

We shall see that this provides us with an inductive formula (from which an explicit formula will be derived) for the rational cohomology of the symplectic quotient of X by K when this exists.

At the end of the last section it was explained how the description of the critical set as the disjoint union of the closed subsets $\{C_\beta \mid \beta \in \mathbf{B}\}$ needs refining in order to state the Morse inequalities of the function f. The problem is that the subsets C_β may be disconnected, and hence the index of the Hessian of f at points of C_β may not be constant. Because of this we defined closed subsets $\{C_{\beta,m} \mid \beta \in \mathbf{B}, 0 \leq m \leq \dim X\}$ such that each C_β is the disjoint union of the subsets $\{C_{\beta,m} \mid 0 \leq m \leq \dim X\}$ and the index of the Hessian of f at any point of $C_{\beta,m}$ is

$$d(\beta,m) = m - \dim K + \dim \mathrm{Stab}\beta$$

(see 4.20).

The statement that the function f is equivariantly perfect for rational coefficients is now equivalent by 2.16 to the equality

5.1.

$$P_t^K(X) = \sum_{\beta,m} t^{d(\beta,m)} P_t^K(C_{\beta,m})$$

where the sum is over all $\beta \in \mathbf{B}$ and $0 \leq m \leq \dim X$.

For each β and m there is a symplectic submanifold $Z_{\beta,m}$ of X acted on by the stabiliser $\mathrm{Stab}\beta$ of β under the adjoint action of K on $\mathbf{k}$ (see 4.19 and 4.23) such that

$$H_K^*(C_{\beta,m};Q) \cong H^*_{\mathrm{Stab}\beta}(Z_{\beta,m} \cap \mu^{-1}(\beta);Q).$$

Thus 5.1 is equivalent to the formula

5.2.

$$P_t^K(X) = \sum_{\beta,m} t^{d(\beta,m)} P_t^{\mathrm{Stab}\beta}(Z_{\beta,m} \cap \mu^{-1}(\beta))$$

where β runs over $\mathbf{B}$ and m over the integers between 0 and dim X.

To show that 5.1 and 5.2 hold, i.e. that the function $f = \|\mu\|^2$ is equivariantly perfect, we shall use the criterion 2.18 together with the following result of Atiyah and Bott (see [A & B] 13.4).

5.3. Suppose that N is a complex vector bundle over a connected space Y and that a compact group K acts as a group of bundle automorphisms of N. Suppose there is a subtorus T_0 of K which acts trivially on Y and that the representation of T_0 on the fibre of N at any point of Y has no nonzero fixed vectors. Then the equivariant Euler class of N in $H_K^*(Y;Q)$ is not a zero-divisor.

5.4. Theorem. Let X be a symplectic manifold acted on by a compact group K with moment map $\mu: X \to \mathbf{k}^*$, and give $\mathbf{k}$ a fixed invariant inner product. Then the function $f = \|\mu\|^2$ on X is equivariantly perfect over the field of rational coefficients. Thus the equivariant Poincaré series of X is given by

$$P_t^K(X) = \sum_{\beta,m} t^{d(\beta,m)} P_t^K(C_{\beta,m})$$

$$= \sum_{\beta,m} t^{d(\beta,m)} P_t^{\mathrm{Stab}\beta}(Z_{\beta,m} \cap \mu^{-1}(\beta)),$$

with each sum running over all elements β of $\mathbf{B}$ and all integers $0 \leq m \leq \dim X$. Here $d(\beta,m) = m - \dim K + \dim \mathrm{Stab}\beta$, and $C_{\beta,m}$, $Z_{\beta,m}$ and $\mathrm{Stab}\beta$ are as defined at 4.19 and 4.9.

Proof. By theorem 4.16 there is a smooth K-invariant stratification

$\{S_\beta \mid \beta \in \mathbf{B}\}$ of X such that for each $\beta \in \mathbf{B}$ the stratum S_β contains the critical subset C_β and the inclusion of C_β in S_β is an equivalence of K-equivariant Cech cohomology.

Let $S_{\beta,m}$ denote the union of those components of S_β which have codimension $d(\beta,m)$. Then by 4.20 $\{S_{\beta,m} \mid \beta \in \mathbf{B}, 0 \leq m \leq \dim X\}$ is a smooth stratification of X such that

$$H_K^*(S_{\beta,m};Q) \cong H_K^*(C_{\beta,m};Q)$$

for each β and m. We must prove that this stratification is equivariantly perfect over the rationals.

By 2.18 it is enough to show that the equivariant Euler class of the normal bundle to each stratum $S_{\beta,m}$ is not a zero-divisor in $H_K^*(S_{\beta,m};Q)$. Under the composition of the isomorphisms

$$H_K^*(S_{\beta,m};Q) \cong H_K^*(C_{\beta,m};Q) \cong H^*_{\mathrm{Stab}\beta}(Z_{\beta,m} \cap \mu^{-1}(\beta);Q)$$

the equivariant Euler class of this normal bundle is identified with the $\mathrm{Stab}\beta$-equivariant Euler class of its restriction, N say, to $Z_{\beta,m} \cap \mu^{-1}(\beta)$.

It follows from 4.18 that the bundle N has a complex structure preserved by the action of $\mathrm{Stab}\beta$. Also from 4.17 we see that N is a quotient of the restriction to $Z_{\beta,m} \cap \mu^{-1}(\beta)$ of the normal bundle to $Z_{\beta,m}$. But by definition $Z_{\beta,m}$ is the union of certain components of the fixed point set of the subtorus T_β of $\mathrm{Stab}\beta$. So by 3.8 the action of T_β on the normal bundle to $Z_{\beta,m}$ has no non-zero fixed vectors. The same is therefore true of the action of T_β on N. Hence by 5.3 the equivariant Euler class of N is not a zero-divisor in $H^*_{\mathrm{Stab}\beta}(Z_{\beta,m} \cap \mu^{-1}(\beta);Q)$. (Note that we should really have considered each component of $Z_{\beta,m}$ separately).

The result follows.

5.5. The subset of X on which the function $f = \|\mu\|^2$ achieves its minimum is $\mu^{-1}(0)$, provided that $\mu^{-1}(0)$ is nonempty. This is a K-invariant subset of X. If we suppose that the stabiliser in K of every $x \in \mu^{-1}(0)$ is finite then the quotient $\mu^{-1}(0)/K$ has a natural symplectic structure and is the symplectic quotient (or Marsden-Weinstein reduction) of X by K.

To see why $\mu^{-1}(0)$ has a natural symplectic structure, note first that if every $x \in \mu^{-1}(0)$ has finite stabiliser then $d\mu(x)$ is surjective for each $x \in \mu^{-1}(0)$. For otherwise there is some $x \in \mu^{-1}(0)$ and some nonzero $a \in \mathbf{k}$ such that

$$0 = d\mu(x)(\xi).a = \omega_x(\xi, a_x)$$

for every $\xi \in T_x X$. (The second equality comes from the definition of a moment map.) Then since ω is nondegenerate $a_x = 0$, so the one-parameter subgroup of K generated by a fixes x, which is impossible.

Thus $\mu^{-1}(0)$ is a submanifold of X, and $\mu^{-1}(0)/K$ is a 'rational homology manifold' (it can be thought of as a manifold except for singularities caused by finite isotropy groups). Moreover from the fact that $\omega_x(\xi, a) = 0$ for $\xi \in T_x\mu^{-1}(0)$ and all $a \in \mathbf{k}$ it is easy to deduce that ω induces a nondegenerate symplectic form on $\mu^{-1}(0)/K$.

In particular if K acts freely on $\mu^{-1}(0)$ then $\mu^{-1}(0)/K$ is a symplectic manifold, and moreover since K is compact the natural map

$$\mu^{-1}(0) \to \mu^{-1}(0)/K$$

is a locally trivial fibration with fibre K. It follows that the natural map

$$\mu^{-1}(0) \times_K EK \to \mu^{-1}(0)/K$$

is a fibration with contractible fibre EK. Hence the equivariant cohomology of $\mu^{-1}(0)$ is isomorphic to the ordinary cohomology of the symplectic quotient of X by K. Moreover for rational cohomology the same is true provided only that the stabiliser of every point in $\mu^{-1}(0)$ is finite. Thus we have

5.6. if the stabiliser of every $x \in \mu^{-1}(0)$ is finite then the rational equivariant cohomology $H_K^*(\mu^{-1}(0);\mathbb{Q})$ of $\mu^{-1}(0)$ is isomorphic to the ordinary rational cohomology $H^*(\mu^{-1}(0)/K;\mathbb{Q})$ of the symplectic quotient $\mu^{-1}(0)/K$.

5.7. Since $\mu^{-1}(0)$ coincides with the critical subset C_0 of X on which the function $f = \|\mu\|^2$ attains its minimum, theorem 5.4 provides a formula for the equivariant Poincaré series of $\mu^{-1}(0)$ in terms of the equivariant Poincaré series of X itself and of all the series

$$P_t^{\mathrm{Stab}\beta}(Z_{\beta,m} \cap \mu^{-1}(\beta))$$

with $\beta \in B$ and $0 \leq m \leq \dim X$. Moreover each $Z_{\beta,m}$ is a compact symplectic manifold on which the compact subgroup $\mathrm{Stab}\beta$ of K acts. We saw at 4.9 that the restriction of μ to Z_β (which is the disjoint union of all the $Z_{\beta,m}$) can be regarded as a moment map for the action of $\mathrm{Stab}\beta$

on Z_β. As usual we use the fixed invariant inner product to identify $\mathbf{k}^*$ with $\mathbf{k}$. Since $\mathrm{Ad}k(\beta) = \beta$ for every $k \in \mathrm{Stab}\beta$ by the definition of $\mathrm{Stab}\beta$, it follows immediately that the map $\mu - \beta$ sending $x \in Z_\beta$ to $\mu(x) - \beta$ is also a moment map for the action of $\mathrm{Stab}\beta$ on Z_β (see the definition of a moment map at 2.3). The same is true when Z_β is replaced by $Z_{\beta,m}$ for any m. As $Z_{\beta,m} \cap \mu^{-1}(\beta)$ is the inverse image of 0 under this moment map, theorem 5.4 will give us an inductive formula for the equivariant cohomology $H_K^*(\mu^{-1}(0);Q)$ of $\mu^{-1}(0)$ provided that we can always calculate $P_t^K(X)$. But for conected groups we have

5.8. Proposition. *Suppose* X *is a compact symplectic manifold acted on by a compact connected Lie group* K *such that a moment map* $\mu: X \to \mathbf{k}^*$ *exists. Then the rational equivariant cohomology of* X *is the tensor product of the ordinary rational cohomology of* X *and that of the classifying space* BK *of* K. *That is*

$$P_t^K(X) = P_t(X)\, P_t(BK) .$$

Remark. If K is not connected let K_0 be its identity component. Then it is not hard to show using 5.8 that $H_K^*(X;Q)$ is the invariant part of $H^*(X;Q) \otimes H^*(BK_0;Q)$ under the action of the finite group K/K_0.

Proof of 5.8. By definition the equivariant cohomology of X is the ordinary cohomology of $X \times_K EK$ where $EK \to BK$ is the classifying bundle for K. Write X_K for $X \times_K EK$.

There is a natural locally trivial fibration $X_K \to BK$ with fibre X. We need to show that this fibration is cohomologically trivial, i.e. that the associated spectral sequence degenerates.

First suppose the group is a torus T. Let β be a generic element of the Lie algebra **t** so that the subgroup $\exp \mathbf{R}\beta$ of T is dense in T, and let $\mu_\beta : X \to \mathbf{R}$ be defined by $\mu_\beta(x) = \mu(x).\beta$. Then by 3.9 μ_β is a nondegenerate Morse function on X and its critical points are the fixed points of T on X. Moreover the induced action of T on the normal bundle to any of the components of the critical set has no nonzero fixed vectors (see 3.8) so by 5.3 it follows that μ_β is equivariantly perfect for T. Thus

$$P_t^T(X) = \sum_C t^{d(C)} P_t(C_T)$$

where C runs over the components of the fixed point set of T and $d(C)$ is the index of μ_β along C. But as T acts trivially on each C we have

$$C_T = C \times_T ET \cong C \times BT,$$

so that

$$P_t(C_T) = P_t(C)\, P_t(BT).$$

Thus

$$P_t^T(X) = P_t(BT) \sum_C t^{d(C)} P_t(C) .$$

The ordinary Morse inequalities for μ_β imply that

$$\sum_C t^{d(C)} P_t(C) - P_t(X) = Q(t)(1 + t)$$

where $Q(t) \geq 0$ in the sense that all its coefficients are nonnegative. In particular

$$P_t(X) \leq \sum_C t^{d(C)} P_t(C) .$$

The Serre spectral sequence for the fibration $X_T \to BT$ starts with

$$E_2^{p,q} = H^p(X;Q) \otimes H^q(BT;Q)$$

and $E_{r+1}^{p,q}$ is the quotient of a subgroup of $E_r^{p,q}$ for each $r \geq 2$. Thus $\dim E_r^{p,q}$ decreases as r increases, so that

$$\dim H^n(X_T;Q) = \sum_{p+q=n} \dim E_\infty^{p,q} \leq \sum_{p+q=n} \dim E_2^{p,q}$$

which implies that

$$P_t^T(X) \leq P_t(X) P_t(BT) .$$

But

$$P_t(X) P_t(BT) \leq P_t(BT) \sum_C t^{d(C)} P_t(C) = P_t^T(X).$$

Therefore both these inequalities must be equalities.

Now let K be any compact connected group with maximal torus T. There are fibrations $BT \to BK$ and $X_T \to X_K$ with fibre the flag manifold K/T. It is well known that the spectral sequences of these fibrations degenerate. To show this one must check that every cohomology class of the fibre K/T extends to a cohomology class of X_T. But the rational cohomology of K/T is multiplicatively generated by the Chern classes of the line bundles L_α on K/T defined by characters α of T. Since $L_\alpha = \mathbf{C} \times_T K$ where the action of T on $\mathbf{C}$ is multiplication by α, the Chern class of the line bundle $\mathbf{C} \times_T (X \times ET)$ over $X_T = X \times_T ET$ restricts to $c_1(L_\alpha)$ on each fibre. Therefore

$$P_t^T(X) = P_t^K(X)\, P_t(K/T)$$

and $P_t(BT) = P_t(BK)\, P_t(K/T)$. The result now follows from the torus case.

5.9. This argument shows that every component μ_β of the moment map is both equivariantly perfect and perfect. The function $f = \|\mu\|^2$ on the other hand is equivariantly perfect by theorem 5.4 but is not necessarily perfect. For example if S^1 acts on the complex sphere as rotation about some axis, then μ is the projection on that axis and has a maximum and a minimum as its only critical points. Thus its Morse series is $1+t^2$ and its equivariant Morse series is $(1+t^2)(1-t^2)^{-1}$. On the other hand $f = \|\mu\|^2$ has critical points at the poles and on the equator, so its equivariant Morse series is

$$1 + 2t^2(1-t^2)^{-1} = (1+t^2)(1-t^2)^{-1}$$

but its ordinary Morse series is $(1+t) + 2t^2$.

The fact that μ_β is perfect is used in work of Carrell and Sommese on $\mathbf{C}^*$ actions on Kåhler manifolds (see [C&S] and also [B-B] and [C&G]).

By using the argument of 5.7 we obtain from theorem 5.4 and proposition 5.8 the inductive formula

5.10.

$$P_t^K(\mu^{-1}(0)) = P_t(X)P_t(BK) - \sum_{\substack{0 \neq \beta \in \mathbf{B} \\ 0 \leq m \leq \dim X}} t^{d(\beta,m)} P_t^{\mathrm{Stab}\beta}(Z_{\beta,m} \cap \mu^{-1}(\beta))$$

for the equivariant cohomology of $\mu^{-1}(0)$ when K is connected.

Moreover if the symplectic quotient exists then by 5.6 its rational cohomology is the same as the rational equivariant cohomology of $\mu^{-1}(0)$, so 5.10 gives us a means of calculating it.

This inductive formula 5.10 was our first goal. It is not hard to deduce from it an explicit formula for $P_t^K(\mu^{-1}(0))$ in terms of the cohomology of certain symplectic submanifolds of X and of the classifying spaces of certain subgroups of K. These submanifolds and subgroups are determined by the combinatorial geometry of the finite set of weights. The remainder of this section will be devoted to obtaining this explicit formula.

As in 3.4 let $\mathbf{A}$ be the set of weights of the action. That is, $\mathbf{A}$ is the image under μ_T of the fixed point set of the maximal torus T of K in X which is a finite set. By definition 3.13 the indexing set $\mathbf{B}$ of the stratification is the set of all minimal weight combinations in the positive Weyl chamber $\mathbf{t}_+$. A minimal weight combination is the closest point to 0 of the convex hull of some nonempty set of weights. Thus any $\beta \in \mathbf{B}$ is the closed point to 0 of $\mathrm{Conv}\{\alpha \in \mathbf{A} \mid (\alpha-\beta).\beta = 0\}$.

We have noted at 5.7 that $Z_\beta \cap \mu^{-1}(\beta)$ is the inverse image of 0 under the map $\mu - \beta: Z_\beta \to \mathrm{Stab}\beta$ and that this is a moment map for the action of $\mathrm{Stab}\beta$ on Z_β. By the definition of Z_β as the union of those components of the fixed point set of T_β on which μ_β takes the value $\|\beta\|^2$, the image under this moment map of the fixed points of T (which is a maximal torus of $\mathrm{Stab}\beta$) on Z_β is just the set

$$\{\alpha - \beta \mid \alpha \in \mathbf{A} \text{ and } (\alpha - \beta).\beta = 0\}.$$

So we make the following definition.

5.11. <u>Definition.</u> A sequence $\beta_1,\ldots,\beta_q$ of nonzero elements of $\mathbf{t}$ is called a β-<u>sequence</u> if for each integer j between 1 and q

(i) β_j is the closest point to 0 of the convex hull

$$\mathrm{Conv}\{\alpha-\beta_1-\beta_2\cdots-\beta_{j-1} \mid \alpha \in \mathbf{A},\ (\alpha-\beta_k).\beta_k = 0 \text{ for } 1\leq k\leq j\}$$

and

(ii) β_j lies in the unique Weyl chamber containing $\mathbf{t}_+$ of the subgroup

$$\bigcap_{1\leq i\leq j} \mathrm{Stab}\beta_i .$$

Note that T is a maximal torus of $\bigcap_{1\leq i\leq j} \mathrm{Stab}\beta_i$ for each j and its Weyl group is a subgroup of the Weyl group of K.

Thus a β-sequence of length 1 is just a nonzero element of the indexing set $\mathbf{B}$, while (β_1,β_2) is a β-sequence of length 2 if and only if $\beta_1 \in \mathbf{B}-\{0\}$ and β_2 lies in the indexing set for the action of $\mathrm{Stab}\beta_1$ on Z_{β_1} with moment map $\mu - \beta_1$.

5.12. Definition. For each β-sequence $\underline{\beta} = (\beta_1,\ldots,\beta_q)$ let $T_{\underline{\beta}}$ be the subtorus of T generated by $\{\beta_1,\ldots,\beta_q\}$; that is, the closure in T of the subgroup generated by the one-parameter subgroups $\{\exp \mathbf{R}\beta_j \mid 1 \leq j \leq q\}$. The fixed point set of $T_{\underline{\beta}}$ on X is a (possibly disconnected) symplectic submanifold of X, and the projection $\mu_{\underline{\beta}}$ of μ onto the Lie algebra of $T_{\underline{\beta}}$ is constant along each of its components. Let $Z_{\underline{\beta}}$ be the union of those components on which $\mu_{\underline{\beta}} = \beta_q$.

5.13. Lemma. If $q \geq 2$ a sequence $\underline{\beta} = (\beta_1,\ldots,\beta_q)$ in $\mathbf{t}$ is a β-sequence if and only if $\beta_1 \in \mathbf{B}-\{0\}$ and the sequence $\underline{\beta}' = (\beta_2,\ldots,\beta_q)$ is a β-sequence for the action of $\mathrm{Stab}\beta_1$ on Z_{β_1} with moment map $\mu - \beta_1$. If this is so then $Z_{\underline{\beta}}$ is contained in Z_{β_1} and coincides with $Z_{\underline{\beta}'}$ where $Z_{\underline{\beta}'}$ is defined relative to the action of $\mathrm{Stab}\beta_1$ on Z_{β_1}.

Proof. This follows directly from the definitions.

If β lies in the Lie algebra $\mathbf{t}_{\underline{\beta}}$ of $T_{\underline{\beta}}$ then every point of Z_β is critical for the function μ_β (see 3.7). Since μ_β is nondegenerate the index $\mathrm{ind}\, H_x(\mu_\beta)$ of the Hessian $H_x(\mu_\beta)$ is constant along connected components of $Z_{\underline{\beta}}$; and so is the index of its restriction to the tangent space of any T_β-invariant submanifold of X containing $Z_{\underline{\beta}}$. So we can make the following definition.

5.14. Definition. Suppose $\underline{\beta} = (\beta_1,\ldots,\beta_q)$ is a β-sequence. For any integer m let $Z_{\underline{\beta},m}$ be the union of those connected components C of $Z_{\underline{\beta}}$ such that if x is any point of C then

$$m = \sum_{1 \leq j \leq q} \text{ind}\, H_x(\mu_{\beta_j}|_{T_j})$$

where $T_j = T_x(Z_{\beta_1} \cap \ldots \cap Z_{\beta_{j-1}})$.

(So $Z_{\underline{\beta},m} = \emptyset$ unless m lies between 0 and $\dim X$).

It follows immediately from the definitions 3.10 and 5.14 that if m and m_1 are any integers then in the terminology of 5.13

5.15 the intersection of $Z_{\underline{\beta},m}$ with Z_{β_1,m_1} is $Z_{\underline{\beta}',m-m_1}$.

Now we can state the explicit formula for $P_t^K(\mu^{-1}(0))$.

5.16. Theorem. Let X be a connected symplectic manifold acted on by a connected compact group K with moment map $\mu: X \to \mathbf{k}^*$, and suppose that $\mathbf{k}$ is equipped with a fixed invariant inner product. Suppose that $\mu^{-1}(0) \neq \emptyset$. Then

$$P_t^K(\mu^{-1}(0)) = P_t(X)\, P_t(BK) + \sum_{\underline{\beta},m} (-1)^q\, t^{d(\underline{\beta},m)}\, P_t(Z_{\underline{\beta},m})\, P_t(B\,\mathrm{Stab}\underline{\beta})$$

the sum being over all β-sequences $\underline{\beta} = (\beta_1,\ldots,\beta_q)$ and all integers $0 \leq m \leq \dim X$. Here β-sequences β and the associated manifolds $Z_{\underline{\beta},m}$ are as defined at 5.11 and 5.14. Also for any β-sequence $\underline{\beta} = (\beta_1,\ldots,\beta_q)$

$$\mathrm{Stab}\underline{\beta} = \bigcap_{1 \leq j \leq q} \mathrm{Stab}\beta_j\,,$$

$B\,\mathrm{Stab}\underline{\beta}$ is the classifying space for $\mathrm{Stab}\underline{\beta}$, and

$$d(\beta,m) = m - \dim K + \dim \operatorname{Stab}\underline{\beta}.$$

Proof. The proof is by induction on $\dim X$. By assumption X is connected and $\mu^{-1}(0) \neq \emptyset$, so that if $\dim X = 0$ then X consists of a single point x and $\mu(x) = 0$. So there are no β-sequences and the result is trivial.

Now assume $\dim X > 0$. By 5.10

(a) $\quad P_t^K(\mu^{-1}(0)) = P_t(X)\,P_t(BX) -$

$$- \sum_{\beta_1, m_1} t^{d(\beta_1,m_1)}\, P_t^{\operatorname{Stab}\beta_1}(Z_{\beta_1,m_1} \cap \mu^{-1}(\beta))$$

where the sum is over nonzero elements β_1 of $\mathbf{B}$ and integers $0 \leq m_1 \leq \dim X$, and $d(\beta_1,m_1) = m_1 - \dim K + \dim \operatorname{Stab} \beta_1$. Moreover $Z_{\beta_1,m_1} \cap \mu^{-1}(\beta_1)$ is the inverse image of 0 under the moment map $\mu - \beta_1$ for the action of $\operatorname{Stab}\beta_1$ on Z_{β_1,m_1}. Recall from 4.8. that because K is connected, so is $\operatorname{Stab}\beta_1$. Without loss of generality we may assume that every component of Z_{β_1,m_1} meets $\mu^{-1}(\beta_1)$. Since $\mu^{-1}(0)$ is nonempty and $\beta_1 \neq 0$, every component of Z_{β_1,m_1} is a proper submanifold of X. Therefore by induction

(b) $\quad P_t^{\operatorname{Stab}\beta_1}(Z_{\beta_1,m_1} \cap \mu^{-1}(\beta)) = P_t(Z_{\beta_1,m_1})\,P_t(B \operatorname{Stab}\beta_1)$

$$+ \sum_{\underline{\beta}',m'} (-1)^{q-1}\, t^{d(\underline{\beta}',m')}\, P_t(Z_{\underline{\beta}',m'})\, P_t(B \operatorname{Stab}\underline{\beta}')$$

where the sum is over β-sequences $\underline{\beta}' = (\beta_2,\ldots\beta_q)$ for the action of

$\mathrm{Stab}\beta_1$ on Z_{β_1} and integers $0 \leq m \leq \dim Z_{\beta_1}$. Moreover

$$\mathrm{Stab}\underline{\beta}' = \bigcap_{2 \leq j \leq q} \mathrm{Stab}\beta_j \cap \mathrm{Stab}\beta_1 = \bigcap_{1 \leq j \leq q} \mathrm{Stab}\beta_j,$$

and $d(\underline{\beta}', m') = m' - \dim \mathrm{Stab}\beta_1 + \dim \mathrm{Stab}\underline{\beta}'$. Therefore the result follows immediately on substituting (b) into (a) and using 5.13 and 5.15.

5.17. Corollary. Under the same assumptions as the theorem suppose that the symplectic quotient of X by K exists. Then its Betti numbers are the same as the equivariant Betti numbers of $\mu^{-1}(0)$, and are thus given by the formula of 5.16.

Proof. This follows immediately from 5.6 and 5.16.

Remark. These results can be extended to the case when K is not connected by using the remark which follows 5.8.

We shall conclude this section with an example.

5.18. Example. As before consider the diagonal action of SU(2) on $(\mathbf{P}_1)^n$. The action of SU(2) on $\mu^{-1}(0)$ is free provided n is odd, since then any configuration with centre of gravity at 0 must contain at least three distinct points. Since SU(1) has rank 1 and β-sequences consist of mutually orthogonal points, every β-sequence must be of length 1 and so can be identified with an element of $\mathbf{B}-\{0\}$. We have seen in 3.17 that any

$\beta \in \mathbf{B}-\{0\}$ corresponds to an integer r such that $\frac{1}{2}n < r \leq n$, and that Z_r consists of sequences containing r copies of 0 and $n-r$ copies of ∞. Thus $Z_{r,m} = \emptyset$ unless $m = 2(r-1)$, and so the rational cohomology of the Marsden-Weinstein reduction is

$$P_t(\mathbf{P}_1^n)\, P_t(BSU(2)) - \sum_{n/2<r\leq n} \binom{n}{r} t^{2(r-1)}\, P_t(BS^1)$$

$$= (1+t^2)^n (1-t^4)^{-1} \sum_{n/2<r\leq n} \binom{n}{r} t^{2(r-1)} (1-t^2)^{-1}\,.$$

When n is odd this is a polynomial in t^2 of degree $n-3$ such that the coefficient of t^{2j} is

$$1 + (n-1) + \binom{n-1}{2} + \dots + \binom{n-1}{\min(j,n-3-j)}\,.$$

It is not a polynomial when n is even.

Further examples will be given in Part II.

§6. Complex group actions on Kähler manifolds

Suppose now that X is a compact Kähler manifold acted on by a complex Lie group G, and that G is the complexification of a maximal compact subgroup K. Thus if **k** and **g** are the Lie algebras of K and G then $\mathbf{g} = \mathbf{k} \oplus i\mathbf{k}$. Suppose also that K preserves the Kähler structure on X. This condition is always satisfied if the Kähler metric is replaced by its average over K. In particular X might be a nonsingular complex projective variety acted on linearly by a complex reductive group (see example 2.1).

The Kähler structure makes X into a symplectic manifold acted on by K, and in addition gives X a K-invariant Riemannian metric. Assume that a moment map $\mu: X \to \mathbf{k}^*$ exists for the action of K on X. This always happens if, for example, K is semisimple, or X is a projective variety, or if $H^1(X;\mathbb{Q}) = 0$. Let $f: X \to \mathbf{R}$ be the norm-square of the moment map with respect to some fixed invariant inner product on **k**.

When applying Morse theory to the function f on a general symplectic manifold we concentrated on the set of critical points for f. We showed that there are Morse inequalities (in fact, equalities) relating the equivariant Betti numbers of X to those of certain critical subsets of f. In order to establish these inequalities, a metric was introduced on X. Then f induced a stratification of X such that the stratum containing any point was determined by the limit of its trajectory under -grad f. This stratification was no more than a technical device: it was not canonically determined by the symplectic group action because it depended on the metric. However in the case of a Kähler manifold there is a canonical choice of metric. We shall see in this section that the stratification induced by f and the Kåhler metric has many elegant properties; in particular, the strata are all complex locally-closed submanifolds of X and are invariant under the action of the complex group G.

6.1. Definition. For $\beta \in \mathbf{B}$ let S_β consist of all points of X whose paths of steepest descent for the Kåhler metric have limit points in the critical subset C_β defined at 3.14.

The subsets $\{S_\beta \mid \beta \in \mathbf{B}\}$ form a stratification of X by lemma 10.7 of the appendix. We shall see that they have the following alternative description in terms of the moment map and the orbits of G.

6.2. A point $x \in X$ lies in S_β if and only if β is the (unique) closest point to 0 of $\mu(\overline{Gx}) \cap \mathbf{t}_+$ (where $\mathbf{t}_+$ is the positive Weyl chamber).

Each stratum S_β also has a decomposition analogous to the decomposition of C_β as $K \times_{\text{Stab}\beta} (Z_\beta \cap \mu^{-1}(\beta))$. It is described as follows. Recall from 4.9 that for each $\beta \in \mathbf{B}$ the symplectic submanifold Z_β of X is acted on by the stabiliser $\text{Stab}\beta$ of β in K and the restriction of $\mu - \beta$ to Z_β is a moment map for the action of $\text{Stab}\beta$ on Z_β.

6.3. Definition. Let Z_β^{min} be the subset of Z_β consisting of those points $x \in Z_\beta$ such that the limit points of the path of steepest descent from x for *the function* $\|\mu - \beta\|^2$ *on* Z_β *lie in* $Z_\beta \cap \mu^{-1}(\beta)$.

So Z_β^{min} is the minimum Morse stratum of Z_β associated to the square of the moment map $\mu - \beta$, and is an open subset of Z_β.

Recall also that Y_β is the subset of X consisting of all those points in X whose paths of steepest descent under μ_β converge to points of Z_β. This subset is a locally closed submanifold of X, and the inclusion of Z_β in Y_β is a cohomology equivalence. In fact since μ_β is nondegenerate (in the sense of Bott) it is straightforward to check that the path of steepest descent of any $y \in Y_\beta$ has a unique limit point, $p_\beta(y)$ say, in Z_β, and that the function $p_\beta : Y_\beta \to Z_\beta$ defined thus is a retraction of Y_β onto Z_β.

6.4. Definition. *Let* Y_β^{min} *be the inverse image of* Z_β^{min} *under the retraction* $p_\beta : Y_\beta \to Z_\beta$. Then Y_β^{min} is an open subset of Y_β and retracts on Z_β^{min}.

We shall see that $S_\beta = GY_\beta^{min}$ for each $\beta \in \mathbf{B}$ and that there is a parabolic subgroup P_β of G which preserves Y_β^{min} such that S_β is isomorphic to $G \times_{P_\beta} Y_\beta^{min}$.

6.5. Example. Suppose $X \subseteq \mathbf{P}_n$ is a complex projective variety acted on linearly by a complex reductive group G, and that $\alpha_0, \ldots, \alpha_n$ are the weights of the representation of G. Then we have seen at 3.11 that

$$Z_\beta = \{(x_0:\ldots:x_n) \in X \mid x_j = 0 \text{ unless } \alpha_j.\beta = \|\beta\|^2\}.$$

It is easily checked that Y_β consists of all $(x_0:\ldots:x_n) \in X$ such that $x_j = 0$ unless $\alpha_j.\beta = \|\beta\|^2$ and $x_j \neq 0$ for some j with $\alpha_j.\beta = \|\beta\|^2$. In particular, suppose $X = (\mathbf{P}_1)^n$ and $G = SL(2)$ acts diagonally on X as in 2.2. By example 3.17 $\mathbf{B}-\{0\}$ can be identified with the set of integers r such that $n/2 < r \leq n$. It is easy to see from 3.17 that $Y_r^{min} = Y_r$ consists of all sequences which contain precisely r copies of 0. But the stratum S_r consists of sequences which contain r coincident points. So $S_r \cong G \times_B Y_r^{min}$ where B is the Borel subgroup of $SL(2)$ fixing 0.

The basic lemma needed is the following.

6.6. Lemma. *If* $\beta \in \mathbf{B}$ *then, for any* $x \in X$,

$$\text{grad}\, \mu_\beta(x) = i\beta_x$$

and

$$\text{grad}\, f(x) = 2i\mu(x)_x,$$

where $\mu(x) \in \mathbf{k}^*$ *is identified with a point of* $\mathbf{k}$ *by using the fixed inner product on* $\mathbf{k}$.

Proof. For any $x \in X$ and $\xi \in T_xX$, if $\langle\, ,\, \rangle$ denotes the Riemannian metric we have

$$\langle \xi, \text{grad}\, \mu_\beta(x) \rangle = d\mu_\beta(x)(\xi) = d\mu(x)(\xi).\beta$$

$$= \omega_x(\xi, \beta_x) = \langle \xi, i\beta_x \rangle .$$

Hence $\operatorname{grad} \mu_\beta(x) = i\beta_x$. The argument used in 3.1. completes the proof.

6.7. Thus the trajectory from any $x \in X$ of $-\operatorname{grad}\mu_\beta$ is

$$\{(\exp -it\beta)x \mid t \geq 0\}.$$

Moreover since $2i\mu(x)$ lies in $\mathbf{g}$, all paths of steepest descent for the function $f = \|\mu\|^2$ on X are contained in orbits of the complex group G.

The complexification T_c of the maximal torus T of K is a maximal complex torus of G.

6.8. Definition. Let B be the Borel subgroup of G associated to the positive Weyl chamber $\mathbf{t}_+$. That is, if

$$\mathbf{g} = \mathbf{t}_c + \sum_\alpha \mathbf{g}^\alpha$$

is the root space decomposition of $\mathbf{g}$ with respect to T_c then $B = \exp \mathbf{b}$ where

$$\mathbf{b} = \mathbf{t}_c + \sum_{\alpha +\mathrm{ve}} \mathbf{g}^\alpha$$

(see [A1] p. 146).

6.9. Lemma. For any $\beta \in \mathbf{t}_+$, let $P_\beta \subseteq G$ consist of all $g \in G$ such that

$(\exp it\beta)g(\exp it\beta)^{-1}$ tends to a limit in G as $t \to -\infty$. Then P_β is a parabolic subgroup of G, and is the product $B\,\mathrm{Stab}\beta$ of the Borel subgroup B with the stabiliser $\mathrm{Stab}\beta$ of β in K.

Proof. It follows from the Peter-Weyl theorem that the compact group K may be embedded in some unitary group $U(n)$. Then as G is the complexification of K it is isomorphic to a subgroup of the complex general linear group $GL(n)$ with Lie algebra $\mathbf{g} = \mathbf{k} + i\mathbf{k} \subseteq \mathbf{g}\ell(n)$. We may assume that the maximal torus T is embedded in the diagonal matrices via $t \to \mathrm{diag}(\alpha_1(t),\ldots,\alpha_n(t))$ where $\alpha_1,\ldots,\alpha_n$ are characters of T. If we identify $\alpha_1,\ldots,\alpha_n$ with elements of $\mathbf{t}^*$ by looking at their derivatives at the identity then β becomes the diagonal matrix with entries $2\pi i(\alpha_j.\beta)$. Without loss of generality suppose that $\alpha_1.\beta \geq \alpha_2.\beta \geq \ldots \geq \alpha_n.\beta$. Then P_β is the subgroup of G which consists of all upper triangular block matrices where the blocks are determined by the different values of $\alpha_i.\beta$.

Given any root α, a matrix $x \in \mathbf{g}$ lies in the root space $\mathbf{g}^\alpha$ if and only if

$$[h,x] = (\alpha.h)x \text{ for all } h \in \mathbf{t}_c .$$

Thus if the (i,j) component of x is nonzero then $\alpha = \alpha_j - \alpha_i$. If moreover α is a positive root then $\alpha.\beta \geq 0$ since $\beta \in \mathbf{t}_+$ so $\alpha_j.\beta \geq \alpha_i.\beta$. This implies that every element of the Lie algebra $\mathbf{b}$ has an upper triangular block decomposition. Hence the same is true of $B = \exp \mathbf{b}$. Thus P_β contains B and so is a parabolic subgroup of G.

Since $G = BK$ (by [A1] p. 147) we deduce that $P_\beta = B(K \cap P_\beta)$. But as $K \subseteq U(n)$, an element of K lies in P_β if and only if it is of block diagonal form; i.e. if and only if it lies in $\mathrm{Stab}\beta$. Therefore $P_\beta = B\,\mathrm{Stab}\beta$ and the proof is complete.

6.10. Lemma. The subsets Y_β and Y_β^{min} of X are invariant under P_β.

Proof. Suppose $p \in P_\beta$ so that $(\exp it\beta)p(\exp it\beta)^{-1}$ tends to some $s \in G$ as $t \to -\infty$. By 6.3 and 6.7 an element y of X lies in Y_β if and only if $(\exp it\beta)y$ converges to an element x of Z_β as $t \to -\infty$. But $(\exp it\beta)y \to x$ as $t \to -\infty$ if and only if $(\exp it\beta)py \to sx$. Clearly s lies in the stabiliser of β in G and hence preserves both Z_β and Z_β^{min}. The result follows.

6.11. Corollary. If $x \in GY_\beta$ then $\|\mu(x)\|^2 \geq \|\beta\|^2$. Equality occurs if and only if $\mu(x)$ lies in the adjoint orbit of β in **k**.

Proof. Since $G = BK$ and $B \subseteq P_\beta$, it follows from 6.10 that $GY_\beta = KY_\beta$. As $\|\mu(kx)\|^2 = \|\mu(x)\|^2$ for all $k \in K$, we can therefore assume that $x \in Y_\beta$. But then the path of steepest descent for the function μ_β from x converges to a point $y \in Z_\beta$ and $\mu_\beta(y) = \|\beta\|^2$ by the definition of Z_β. So

$$\mu(x).\beta = \mu_\beta(x) \geq \mu_\beta(y) = \|\beta\|^2 ,$$

from which the result follows immediately.

6.12. Corollary. *If* $x \in GY_\beta^{min}$ *then* β *is the unique closest point to* 0 *of* $\mu(\overline{Gx}) \cap \mathbf{t}_+$.

Proof. Since the adjoint orbit of β in $\mathbf{k}$ intersects $\mathbf{t}_+$ only at β, by 6.11 it suffices to show that β lies in $\mu(\overline{Gx})$. Without loss of generality $x \in Y_\beta^{min}$ so that $(\exp it\beta)x$ converges to some $y \in Z_\beta^{min}$ as $t \to -\infty$. Then $\overline{Gy} \subseteq \overline{Gx}$, so it is enough to show that $\beta \in \mu(\overline{Gy})$.

But by the definition of Z_β^{min} the path of steepest descent from y of the function $\|\mu - \beta\|^2$ restricted to Z_β has a limit point in $Z_\beta \cap \mu^{-1}(\beta)$. By 4.9 $\mu - \beta$ is a moment map for the action of $\mathrm{Stab}\beta$ on Z_β so 6.7 implies that this path is contained in the orbit of y under the complexification of $\mathrm{Stab}\beta$. So $\beta \in \mu(\overline{Gy})$. This completes the proof.

What we are aiming to do is to show that $GY_\beta^{min} = S_\beta$ for each $\beta \in \mathbf{B}$. Then 6.12 will show that 6.2 is as claimed an alternative definition of the stratification $\{S_\beta \mid \beta \in \mathbf{B}\}$.

6.13. Remark. The definition 6.2 is neater but less useful, and moreover cannot be given directly without some guarantee that $\mu(\overline{Gx}) \cap \mathbf{t}_+$ contains a unique point which is closest to 0. In fact it has been proved recently by Mumford that if X is a complex projective variety acted on linearly by G

then $\mu(\overline{Gx}) \cap t_+$ is convex for every $x \in X$, which implies that it contains a unique point closest to 0 (see the appendix to [Ne]). Indeed the same is true when $\overline{Gx}$ is replaced by any closed G-invariant subvariety of X. This generalises a similar result of Guillemin and Sternberg which requires the subvariety to be nonsingular and hence does not apply to $\overline{Gx}$ in general. Mumford's proof is algebraic, but it can be adapted to the more general Kåhler case by using lemma 8.8 and remark 8.9 below.

The next result we need is that GY_β^{min} is diffeomorphic to $G \times_{P_\beta} Y_\beta^{min}$. This will imply in particular that GY_β^{min} is smooth.

We shall first make the following assumption.

6.14. Assumption. The minimum stratum X^{min} for the function $\|\mu\|^2$ on X is contained in the minimum stratum denoted by X_T^{min} for the function $\|\mu_T\|^2$. Here as before μ_T is the composition $X \overset{\mu}{\rightarrow} k^* \rightarrow t^*$ and is a moment map for the action of the maximal torus T on X.

The proofs of the next lemma and theorem 6.18 will depend on this assumption holding for all actions of closed subgroups of K. But clearly the assumption is valid for all tori, so that theorem 6.18 will hold for the torus T at least; and from this we will be able to deduce that the assumption is always valid.

6.15. Lemma. *If* $x \in Y_\beta^{min}$ *then* $\{g \in G \mid gx \in Y_\beta^{min}\} = P_\beta$ *and*

$\{a \in \mathbf{g} | a_x \in T_x Y_\beta^{min}\} = \mathbf{p}_\beta$ <u>where</u> $\mathbf{p}_\beta$ <u>is the Lie algebra of</u> P_β .

<u>Proof.</u> Lemma 6.10 shows that $P_\beta \subseteq \{g \in G | gx \in Y_\beta^{min}\}$. For the reverse inclusion suppose that $g \in G$ is such that $gx \in Y_\beta^{min}$. Let $N_K(T)$ be the normaliser of T in K. Then $G = BN_K(T)B$ by the Bruhat decomposition (see e.g. [A1] p. 135), so that

$$g = b_1 k b_2$$

with $k \in N_K(T)$ and $b_1, b_2 \in B$. Since $B \subseteq P_\beta$ and both x and gx lie in Y_β^{min}, so do $b_2 x$ and $kb_2 x = b_1^{-1}(gx)$. By assumption 6.14 applied to the moment map $\mu - \beta$ for the action of $\mathrm{Stab}\beta$ on Z_β we have $Z_\beta^{min} \subseteq Z_{\beta,T}^{min}$. Therefore by applying 6.12 to the action of the complex torus T_c on X it follows that β is the closest point to 0 in $\mathbf{t}$ of both $\mu_T(\overline{T_c b_2 x})$ and $\mu_T(\overline{T_c k b_2 x})$. Since k normalises T and T_c we have that

$$\mu_T(\overline{T_c k b_2 x}) = \mu_T(\overline{k T_c b_2 x}) = \mathrm{Ad}k\, \mu_T(\overline{T_c b_2 x}).$$

As the inner product on $\mathbf{k}$ is invariant under the adjoint action, this implies that $\mathrm{Ad}k(\beta) = \beta$, i.e. that $k \in \mathrm{Stab}\beta$. Since $P_\beta = B\,\mathrm{Stab}\beta$ by 6.9, the element $g = b_1 k b_2$ of G therefore lies in P_β .

It remains to show that $\{a \in \mathbf{g} | a_x \in T_x Y_\beta^{min}\} \subseteq \mathbf{p}_\beta$ since we know the reverse inclusion holds. By 6.9 $\mathbf{p}_\beta = \mathbf{b} + \mathrm{stab}\beta$, so it suffices to show that

(a) $$\{a \in \mathbf{k} | a_x \in T_x Y_\beta^{min}\} \subseteq \mathrm{stab}\beta .$$

This has already been proved (see lemma 4.10) in the particular case when $x \in Z_\beta \cap \mu^{-1}(\beta)$. Moreover (a) is a linear independence condition on x, and hence the subset of Y_β^{min} where it holds is an open neighbourhood of

$Z_\beta \cap \mu^{-1}(\beta)$. It is also clearly invariant under P_β.

But by the proof of 6.12, given any $x \in Y_\beta^{min}$ there is some $y \in Z_\beta \cap \mu^{-1}(\beta)$ which lies in the closure of the orbit of x under the complexification $Stab_c \beta$ of $Stab\beta$. Since $Stab_c \beta \subseteq P_\beta$ it follows that the only P_β-invariant neighbourhood of $Z_\beta \cap \mu^{-1}(\beta)$ is Y_β^{min} itself. Thus (a) must hold for every $x \in Y_\beta^{min}$. This completes the proof.

It follows from this lemma, by adapting the argument of corollary 4.11, that

6.16. GY_β^{min} <u>is smooth and is diffeomorphic to</u> $G \times_{P_\beta} Y_\beta^{min}$.

6.17. <u>Remark.</u> In fact Y_β is a locally-closed complex submanifold of X, from which it follows immediately that GY_β^{min} is also complex and is isomorphic as a complex manifold to $G \times_{P_\beta} Y_\beta^{min}$.

To see that Y_β is complex, recall that by definition Y_β consists of those points $y \in X$ whose trajectories under $-\mathrm{grad}\mu_\beta$ converge to points of Z_β. Since $\mathrm{grad}\mu_\beta(x) = i\beta_x$ for all $x \in X$ the vector field $-\mathrm{grad}\mu_\beta$ on X is holomorphic. Moreover by the proof of 4.12 if $x \in Z_\beta$ then the Hessian $H_x(\mu_\beta)$ acts as a complex-linear transformation of the tangent space T_xX, and depends holomorphically on x (again because the action of the group is complex-analytic). Since μ_β is nondegenerate the local theory of ordinary differential equations tells us that Y_β is a complex submanifold of X in some neighbourhood of Z_β (see e.g. the details of [H] IX 5). But for every $y \in Y_\beta$ there is some $t \in \mathbf{R}$ such that the point $(\exp it\beta)y$ of the path of steepest descent for μ_β from y lies in this neighbourhood of Z_β. Since $\exp it\beta$ acts as a complex analytic isomorphism of X which preserves Y_β it follows that Y_β is a complex submanifold of X as required.

We can now prove the result we want, on the assumption that 6.14 and hence also 6.16 hold.

6.18. Theorem. Suppose X is a compact Kåhler manifold acted on by a complex Lie group G and that G is the complexification of a maximal compact subgroup K which preserves the Kåhler structure. Suppose that $\mu: X \to \mathbf{k}^*$ is a moment map for the action of K on X and let $\{S_\beta \mid \beta \in \mathbf{B}\}$ be the Morse stratification for the function $f = \|\mu\|^2$ on X (see 6.1). Then for each $\beta \in \mathbf{B}$ we have $S_\beta = GY_\beta^{min}$, and $x \in S_\beta$ if and only if β is the unique closest point to 0 of $\mu(\overline{Gx}) \cap \mathbf{t}_+$. The subsets $\{S_\beta \mid \beta \in \mathbf{B}\}$ form a smooth stratification of X which is G-equivariantly perfect. Moreover $S_\beta \cong G \times_{P_\beta} Y_\beta^{min}$ so that $H_G^*(S_\beta;Q) \cong H_{P_\beta}^*(Y_\beta^{min};Q)$ for each $\beta \in \mathbf{B}$.

Proof. First we shall use the results of the appendix to show that $S_\beta = GY_\beta^{min}$.

Since Y_β^{min} is invariant under the action of the parabolic subgroup P_β we have that $GY_\beta^{min} = KY_\beta^{min}$. So proposition 4.15 implies that some open subset of GY_β^{min} is a minimising manifold for the function $f = \|\mu\|^2$ along C_β. Moreover by 6.7 the trajectory under $-\text{grad}f$ of any $x \in GY_\beta^{min}$ is contained in the orbit Gx and hence in GY_β^{min}. In particular the gradient flow of f is tangential to GY_β^{min}, so by theorem 10.4 of the appendix GY_β^{min} coincides with the stratum S_β in some neighbourhood U of the critical subset C_β.

Suppose $x \in S_\beta$. Then the path of steepest descent for $f = \|\mu\|^2$ from x has a limit point in C_β and therefore intersects $U \cap S_\beta = U \cap GY_\beta^{min}$. But by 6.7 this path is contained in the orbit Gx so x must lie in GY_β^{min}. Thus $S_\beta \subseteq GY_\beta^{min}$ for each $\beta \in \mathbf{B}$. But 6.12 implies that the subsets $\{GY_\beta^{min} \mid \beta \in \mathbf{B}\}$ are disjoint. Since $X = \bigcup_{\beta \in \mathbf{B}} S_\beta$ it follows that $S_\beta = GY_\beta^{min}$ for each $\beta \in \mathbf{B}$.

We have already seen at 6.16 that $GY_\beta^{min} \cong G \times_{P_\beta} Y_\beta^{min}$ for each β, and hence $H_G^*(GY_\beta^{min};Q) \cong H_{P_\beta}^*(Y_\beta^{min};Q)$ by [A&B]§13. Finally theorem 5.4 shows that the stratification $\{S_\beta \mid \beta \in \mathbf{B}\}$ is equivariantly perfect for K, and this implies immediately that it is equivariantly perfect for G since K and G are homotopically equivalent. (Alternatively the proof of 5.4 can be adapted easily to work for the complex group).

It remains to check that the assumption 6.14 is valid.

Notation. If it is necessary to make clear what group is involved, the stratum S_β will be written as $S_{\beta,K}$.

6.19. Lemma. Assumption 6.14 is always valid. That is, the minimum stratum $X^{min} = S_{0,K}$ is contained in the minimum stratum $X_T^{min} = S_{0,T}$ associated to the action of the maximal torus T of K.

Proof. Since 6.14 holds trivially for tori, proposition 6.18 is valid for the maximal torus T of K. Thus if x does not lie in the minimum stratum $X_T^{min} = S_{0,T}$ for the torus then there exists some nonzero $\beta \in \mathbf{B}$ such that $x \in S_{\beta,T} \subseteq T_c Y_\beta$. (Note that Y_β is the same whether the group is K or T; also $T_c Y_\beta = Y_\beta$). Thus by corollary 6.11 if $y \in \overline{Gx}$ then $\|\mu(y)\|^2 \geq \|\beta\|^2 > 0$. Since the path of steepest descent for the function $\|\mu\|^2$ from x is contained in Gx, we deduce that x cannot lie in X^{min}.

The proof of theorem 6.18 for any group, torus or not, is now complete.

6.20. Remark. By theorem 4.16 the inclusion of the minimum set $\mu^{-1}(0)$ for f in the minimum stratum X^{min} is an equivalence of equivariant cohomology. So 5.10 and 5.16 may be interpreted as formulae for the equivariant Poincaré series $P_t^G(X^{min})$. These formulae can also be derived directly from theorem 6.18.

If G acts freely on the open subset X^{min} of X then the quotient X^{min}/G is a complex manifold, and it would be natural to hope that the rational cohomology of this is isomorphic to $H_G^*(X^{min};Q)$. This could be proved by showing that the quotient map $X^{min} \to X^{min}/G$ is a locally trivial fibration. However this is unnecessary because in the next section we shall see that X^{min}/G is homeomorphic to the symplectic quotient $\mu^{-1}(0)/K$. This reduces the problem to the action of a compact group.

Let us conclude this section by considering how the stratification is affected if we alter the choice of moment map or of the invariant inner product on **k**. (From the algebraic point of view, changing the moment map on a complex projective variety X corresponds to changing the projective embedding of X).

First consider the inner product. Clearly if the group is a torus then any inner product is invariant, and different choices give different stratifications. For an example take $(\mathbf{C}^*)^2$ acting on $\mathbf{P}_1$ via the map $\phi: (\mathbf{C}^*)^2 \to GL(2)$ given by

$$\phi(h) = \begin{bmatrix} \alpha_0(h) & 0 \\ 0 & \alpha_1(h) \end{bmatrix}$$

where $\alpha_i: (\mathbf{C}^*)^2 \to \mathbf{C}^*$ is the projection onto the $(i+1)$th factor. Then the stratum to which an element $(x_0:x_1) \in \mathbf{P}_1$ belongs is determined by the closest point to 0 in the convex hull of $\{\alpha_i \mid x_i \neq 0\}$. But α_0 and α_1 are linearly independent so there exist inner products on the Lie algebra of the torus for which the closest point to 0 of their convex hull is respectively α_0, α_1, and neither of these. These give different stratifications of $\mathbf{P}_1$.

On the other hand if G is semisimple then the stratification is independent of the choice of inner product. For then G is, up to finite central extensions, the product $G_1 \times \ldots \times G_k$ of simple groups G_i with maximal compact subgroups K_i and maximal compact subtori T_i say. Then $\mathbf{k} = \mathbf{k}_1 \oplus \ldots \oplus \mathbf{k}_k$ and $\mathbf{k}_1, \ldots, \mathbf{k}_k$ must be mutually orthogonal under any invariant inner product on $\mathbf{k}$. For each i the projection μ_i of μ onto $\mathbf{k}_i$ is a moment map for the action of K_i on X. It is not hard to see that for any $x \in X$ the closest point to 0 of $\mu(\overline{Gx}) \cap \mathbf{t}_+$ is

$$\beta = \beta_1 + \ldots + \beta_k$$

where β_i is the closest point to 0 of $\mu_i(\overline{G_i x}) \cap (\mathbf{t}_i)_+$. But because $\mathbf{k}_i$ is simple the invariant inner product on $\mathbf{k}_i$ is unique up to scalar multiplication, and therefore each β_i is independent of the choice of inner product.

Now consider the effect of changing the choice of moment map μ. The only possible way to do this is to add to μ a constant $\xi \in \mathbf{k}^*$ which is invariant under the adjoint action. Thus as has already been noted, when G is semisimple the moment map is unique. On the other hand if G is a torus T_c an arbitrary constant may be added to the moment map. We know that the stratum containing any point x is labelled by the closest point to 0 of $\mu(\overline{T_c x})$ which is the convex hull of some subset of $\{\alpha_0, \ldots, \alpha_n\}$. Thus by adding different constants to μ one can obtain a finite number of distinct stratifications of X.

Since any compact group is, up to finite central extensions, the product of a torus by a semisimple group it is now easy to deduce what happens in general.

§7. Quotients of Kähler manifolds

Suppose as in §6 that X is a compact Kähler manifold acted on by a complex Lie group G, and that G is the complexification of a maximal compact subgroup K. Assume also that K preserves the Kähler form ω on X (if necessary replace ω by its average over K), and that a moment map $\mu: X \to \mathbf{k}^*$ exists for the symplectic action of K on X.

Any torus in G will always have fixed points in X so we cannot hope to give the topological quotient X/G the structure of a Kähler manifold. However in good cases there is a compact Kähler manifold which it is natural to regard as the 'Kähler quotient' of the action of G on X. When X is a complex projective variety on which G acts linearly, this quotient coincides with the projective quotient defined by Mumford using geometric invariant theory. The good cases occur when the stabiliser in K of every $x \in \mu^{-1}(0)$ is finite. Recall that this is the condition needed for there to be a symplectic quotient associated to the action.

As before let X^{min} be the subset of X consisting of points whose paths of steepest descent under the function $f = \|\mu\|^2$ have limit points in $\mu^{-1}(0)$. By 6.18 X^{min} is a G-invariant open subset of X. We shall see that when K acts with finite stabilisers on $\mu^{-1}(0)$ then the symplectic quotient $\mu^{-1}(0)/K$ can be identified with X^{min}/G and thus has a complex structure. The symplectic form induced on $\mu^{-1}(0)/K$ is then holomorphic and makes $\mu^{-1}(0)/K$ into a compact Kähler manifold except for the singularities caused by finite isotropy groups. (Manifolds with such singularities have been well studied; they are sometimes called V-manifolds). This is the natural Kähler quotient of X by G.

The rational cohomology of this quotient can be calculated by using 5.10 or 5.17.

Recall from 5.5 that the condition that K acts with finite stabilisers on $\mu^{-1}(0)$ implies that $\mu^{-1}(0)$ is smooth. The inclusion of $\mu^{-1}(0)$ in X^{min} induces a natural continuous map

$$\mu^{-1}(0)/K \to X^{min}/G.$$

In order to show that this map is a homeomorphism we need some lemmas. The first is

7.1. Lemma. $G = K \exp i\mathbf{k}$.

Proof. The left coset space G/K is a complete Riemannian manifold (see [He]), so that the associated exponential map $\mathrm{Exp}: T_K(G/K) \to G/K$ maps the tangent space at the coset K onto G/K. Moreover $T_K(G/K) = \mathbf{g}/\mathbf{k}$ and by [He] p. 169(4) we have

$$\mathrm{Exp}(a+\mathbf{k}) = (\exp a)K \text{ for any } a \in \mathbf{g}.$$

Since $\mathbf{g} = \mathbf{k} + i\mathbf{k}$ the result follows.

Next we need

7.2. Lemma. If $x \in \mu^{-1}(0)$ then $Gx \cap \mu^{-1}(0) = Kx$.

<u>Proof.</u> Suppose $g \in G$ is such that $gx \in \mu^{-1}(0)$. We wish to show that there exists $k \in K$ such that $gx = kx$. Since $\mu^{-1}(0)$ is K-invariant, by 7.1 it suffices to consider the case $g = \exp ia$ where $a \in \mathbf{k}$.

Let $h: \mathbf{R} \to \mathbf{R}$ be defined by $h(t) = \mu((\exp iat)x).a$. Then h vanishes at 0 and 1 because x and $(\exp ia)x$ both lie in $\mu^{-1}(0)$. Therefore there is some $t \in (0,1)$ such that

$$0 = h'(t) = d\mu(y)(ia_y).a = \omega_y(ia_y, a_y) = \langle a_y, a_y \rangle$$

where $y = (\exp iat)x$ and $\langle \, , \, \rangle$ denotes the metric induced by the Kåhler structure. Thus $a_y = 0$, so that $\exp ia\mathbf{R}$ fixes y and hence also x. But then $(\exp ia)x = x \in Kx$, so the proof is complete.

It is necessary to strengthen this result.

7.3. <u>Lemma. Suppose</u> x <u>and</u> y <u>lie in</u> $\mu^{-1}(0)$ <u>and</u> $x \notin Ky$. <u>Then there exist disjoint G-invariant neighbourhoods of</u> x <u>and</u> y <u>in</u> X.

<u>Proof.</u> Since K is compact and $x \notin Ky$ there is a compact K-invariant neighbourhood V of x in $\mu^{-1}(0)$ not containing y. Since $G = (\exp i\mathbf{k})K$ by 7.1, it suffices to show that $(\exp i\mathbf{k})V$ is a neighbourhood of x in X and that $y \notin \overline{(\exp i\mathbf{k})V}$.

To see that $(\exp i\mathbf{k})V$ is a neighbourhood of x in X consider the map $\sigma: \mathbf{k} \times \mu^{-1}(0) \to X$ which sends (a,w) to $(\exp ia)w$. This is a smooth map of smooth manifolds, so it is enough to show that its derivative at $(1,x)$ is surjective. If not, there exists some nonzero $\xi \in T_xX$ such that

$<\xi,\zeta> = 0$ for all ζ in the image of $d\sigma(1,x)$. In particular

$$<\xi, ia_x> = 0$$

for all $a \in \mathbf{k}$. But then

$$0 = \omega_x(\xi, a_x) = d\mu(x)(\xi).a$$

for all $a \in \mathbf{k}$. Thus

$$\xi \in \ker d\mu(x) = T_x(\mu^{-1}(0))$$

and hence ξ lies in the image of $d\sigma(1,x)$, which is a contradiction.

Therefore if

$$W = \exp\{ia \mid a \in \mathbf{k}, \|a\| \leq 1\}V$$

then W is a compact neighbourhood of x in X. Let

$$\varepsilon = \inf\{<a_w, a_w> \mid w \in W, a \in \mathbf{k}, \|a\| = 1\}.$$

If $w \in W$ then w lies in the G-orbit of some $z \in \mu^{-1}(0)$ and it follows easily from the proof of 7.2 that the stabiliser of w in G is finite. Therefore $a_w \neq 0$ whenever $0 \neq a \in \mathbf{k}$, and so $\varepsilon > 0$.

Now suppose z lies in V and $a \in \mathbf{k}$ is such that $\|a\| = 1$. Consider the function $h: \mathbf{R} \to \mathbf{R}$ given by $h(t) = \mu((\exp ita)z).a$. As in the proof of 7.2, if $t \in \mathbf{R}$ then $h'(t) = <a_w, a_w>$ where $w = (\exp ita)z$. Therefore $h'(t) \geq 0$ for all $t \in \mathbf{R}$, and $h'(t) \geq \varepsilon$ when $t \in [0,1]$ by the choice of ε. Since $h(0) = 0$ the mean value theorem implies that $h(t) \geq \varepsilon$ when $t \geq 1$. As $\|a\| = 1$ it follows that $\|\mu(\exp ita)z\| \geq \varepsilon$ when $t \geq 1$.

We deduce that if $z \in V$ then $\|\mu(\exp iaz)\| \geq \varepsilon$ whenever $a \in \mathbf{k}$ and $\|a\| \geq 1$. Hence as V is compact $(\exp i\mathbf{k})V$ is closed in a neighbourhood of $\mu^{-1}(0)$. Since $y \in \mu^{-1}(0)$ and $y \notin (\exp i\mathbf{k})V$ by 7.2, it follows that

$y \notin \overline{(\exp i\mathbf{k})V}$.

Now we can prove the result we were aiming for.

7.4. Theorem. Let X be a Kåhler manifold acted on by a group G which is the complexification of a maximal compact subgroup K that preserves the Kåhler structure on X. Suppose that a moment map $\mu: X \to \mathbf{k}^*$ exists for this action of K and suppose that the stabiliser in K of every $x \in \mu^{-1}(0)$ is finite. Then $X^{min} = G\mu^{-1}(0)$ and the natural map $\mu^{-1}(0)/K \to X^{min}/G$ is a homeomorphism.

Proof. $G\mu^{-1}(0) \subseteq X^{min}$ because X^{min} is G-invariant by 6.18 and contains $\mu^{-1}(0)$. Conversely if $x \in X^{min}$ then there is some $y \in \mu^{-1}(0)$ lying in the closure of the path of steepest descent for $\|\mu\|^2$ from x. By 6.7 this path is contained in the orbit Gx, so that $y \in \overline{Gx}$. Then $Gy \subseteq \overline{Gx}$, so that either $y \in Gx$ or $\dim Gy < \dim Gx$. But by assumption the stabiliser of y in K is finite, and this implies that $\dim Gy = \dim G \geq \dim Gx$ (see 7.2). We conclude that $y \in Gx$, so that $x \in G\mu^{-1}(0)$.

Thus $X^{min} = G\mu^{-1}(0)$, so the natural map $\mu^{-1}(0)/K \to X^{min}/G$ is surjective. Lemma 7.2 implies that it is injective, while lemma 7.3 shows that X^{min}/G is a Hausdorff space. Thus the map is a continuous bijection from a compact space to a Hausdorff space, and therefore it is a homeomorphism.

It follows easily from the proof of 7.2 that if K acts freely on $\mu^{-1}(0)$ then G acts freely on the open subset X^{min} of X, so that the complex structure on X^{min} induces a complex structure on the topological quotient $X^{min}/G = \mu^{-1}(0)/K$. The symplectic form on $\mu^{-1}(0)/K$ induced by ω is holomorphic with respect to this complex structure because ω is holomorphic on X, and indeed is a Kåhler form because ω is Kåhler. Hence the quotient $X^{min}/G = \mu^{-1}(0)/K$ is a compact Kåhler manifold. More generally when the stabiliser of every point in $\mu^{-1}(0)$ is finite the quotient X^{min}/G $= \mu^{-1}(0)/K$ can be thought of as a Kåhler manifold with singularities caused by the finite isotropy groups.

7.5. Remark. The proof of lemma 7.2 is independent of the assumption that the stabiliser of every point in $\mu^{-1}(0)$ is finite, and it is also possible to prove lemma 7.3 without using this assumption. One uses the fact that if $a \in \mathbf{k}$ then the function μ_a defined by $\mu_a(x) = \mu(x).a$ is a nondegenerate Morse function on X. This implies that given any point $y \in \mu_a^{-1}(0)$ and any neighbourhood U of y in X, there is a smaller neighborhood V of y and $\varepsilon > 0$ such that the intersection with $\mu_a^{-1}[-\varepsilon,\varepsilon]$ of any trajectory of grad μ_a which passes through a point of V is contained in U. (The proof of this when y is not critical for μ_a is easy: see the proof of 7.3). Using this the argument of 7.3 gives the result when $G = \mathbf{C}^*$, and the torus case also follows without difficulty. The general case can then be deduced from the facts that $G = KT_cK$ and that K is compact.

From this it follows without the assumption of finite stabilisers that any $x \in X$ lies in $G\mu^{-1}(0)$ if and only if x lies in X^{min} and its orbit Gx is closed in X^{min}; and also that the natural map $\mu^{-1}(0)/K \to G\mu^{-1}(0)/G$ is a homeomorphism. In particular when X is a projective variety on which G acts linearly one finds that $\mu^{-1}(0)/K$ is naturally homeomorphic to the geometric invariant theory quotient of X by G.

§8. The relationship with geometric invariant theory

From now on we shall assume that our Kähler manifold X is in fact a nonsingular complex projective variety and that G is a connected reductive complex group acting linearly on X as in example 2.1. Then geometric invariant theory associates to the action of G on X a projective 'quotient' variety M (see [M]). In fact M is the projective variety $\operatorname{Proj} A(X)^G$ where $A(X)^G$ is the invariant subring of the coordinate ring of X. In general M has bad singularities even though X is nonsingular. However in good cases M coincides with the quotient in the usual sense of an open subset X^{ss} of X by G and the stabiliser in G of every $x \in X^{ss}$ is finite. This implies that M behaves like a manifold for rational cohomology.

It turns out that the geometric invariant theory quotient M coincides with the symplectic quotient $\mu^{-1}(0)/K$, and that the good cases occur precisely when the stabiliser in K of every $x \in \mu^{-1}(0)$ is finite. So the work of the preceding sections can be used to obtain formulae for the Betti numbers of M in these cases. The formulae involve the cohomology of X and various subvarieties, together with that of the classifying space of G and certain reductive subgroups.

8.1. Remark. The example of $PGL(n+1)$ shows that the assumption that G acts on X via a homomorphism $\phi: G \to GL(n+1)$ involves some loss of generality. However the finite cover $SL(n+1)$ of $PGL(n+1)$ has the same

Lie algebra, moment map and orbits on X as PGL(n+1). Moreover if G is a connected reductive linear algebraic group acting algebraically on a smooth projective variety $X \subseteq \mathbf{P}_n$ then the action is given by a homomorphism $\phi: G \to PGL(n+1)$, provided we assume that X is not contained in any hyperplane. The argument for this runs as follows. First we note that the induced action of G on the Picard variety Pic(X) of X is trivial. For it is enough to show that every Borel subgroup B of G acts trivially. But by [B] theorem 10.4 B has a fixed point on each component of Pic(X). Applying this with X replaced by Pic(X) we see that there is an ample bundle on Pic(X) fixed by B. By the theorem of [G & H] p. 326 it follows that the image of B in the group of automorphisms of Pic(X) is discrete. Thus as B is connected it must act trivially. (Alternatively see [M] corollary 1.6). Now let L be the hyperplane bundle on $X \subseteq \mathbf{P}_n$, which has automorphism group GL(n+1). Then $g^*L \cong L$ for all $g \in G$, so that the action of any g on X is covered by an automorphism of L and hence is given by an element of PGL(n+1). This element is unique because X is not contained in a hyperplane. So we get a well-defined homomorphism $\phi: G \to PGL(n+1)$ which induces the action of G on X.

We may now replace G by its image in PGL(n+1) and then by the inverse image of this in SL(n+1) to obtain a linear action on X with essentially the same properties as the original action.

The inclusion of $A(X)^G$ in $A(X)$ induces a surjective G-invariant morphism $\psi: X^{ss} \to M$ from an open subset X^{ss} of X to the quotient M. We shall see that X^{ss} always coincides with the minimum Morse stratum X^{min} associated to the function $f = \|\mu\|^2$ on X. Therefore §5 and §6 give us formulae for the equivariant Betti numbers of X^{ss}. It may happen that a fibre of ψ contains more than one orbit of G, so that $M \neq X^{ss}/G$. However there is an open subset X^s of X^{ss} such that every fibre which meets X^s is a single G-orbit (see [M] theorem 1.10). The image of X^s in M is an open subset M' of M and $M' = X^s/G$.

8.2. Definitions (see [M] definitions 1.7 and 1.8, noting that Mumford calls stable points "properly stable": this seems to be no longer the accepted terminology). A point $x \in X$ is semistable if there is a homogeneous non-constant polynomial $F \in \mathbf{C}[X_0,\ldots,X_n]$ which is invariant under the natural action of G on $\mathbf{C}[X_0,\ldots,X_n]$ and is such that $F(x) \neq 0$. x is stable if there is an invariant F with $F(x) \neq 0$ such that all orbits of G in the affine set

$$X_F = \{y \in X \mid F(y) \neq 0\}$$

are closed in X_F and in addition the stabiliser of x in G is finite.

X^{ss} is the set of semistable points of X and X^s is the set of stable points.

8.3. Remark. Suppose that the stabiliser in G of every semistable point in

X is finite. Then if $x \in X^{ss}$ there exists some homogeneous non-constant G-invariant polynomial F such that $F(x) \neq 0$. Every point in $X_F = \{y \in X \mid F(y) \neq 0\}$ is semistable, so every G-orbit in X_F has the same dimension as G. This implies that every orbit is closed in X_F and thus that x is stable. Hence $X^s = X^{ss}$.

We shall use the following facts which follow from [M] theorem 2.1 and proposition 2.2.

8.4. A point $x \in X$ is semistable for the action of G on X if and only if it is semistable for the action of every 1-PS (one-parameter subgroup) $\lambda: \mathbf{C}^* \to G$ of G on X.

8.5. If $\lambda: \mathbf{C}^* \to GL(n+1)$ is given by

$$z \to \mathrm{diag}(z^{r_0},\ldots,z^{r_n})$$

with $r_0,\ldots,r_n \in \mathbf{Z}$, then a point $x = (x_0:\ldots:x_n) \in \mathbf{P}_n$ is semistable for the action of $\mathbf{C}^*$ via λ if and only if

$$\min\{r_j \mid x_j \neq 0\} \leq 0 \leq \max\{r_j \mid x_j \neq 0\}.$$

Using this last fact, we obtain

8.6. Lemma. <u>When $G = \mathbf{C}^*$ the set X^{ss} of semistable points coincides with the minimum Morse stratum X^{min} associated to the function $\|\mu\|^2$.</u>

Proof. There are coordinates in $\mathbf{P}_n$ such that $G = \mathbf{C}^*$ acts diagonally by

$$z \to \operatorname{diag}(z^{r_0},\ldots,z^{r_n})$$

say. We have $K = \{e^{2\pi it} \mid t \in \mathbf{R}\}$ so that $\phi_*(\mathbf{k})$ is the subspace of $\mathbf{u}(n+1)$ spanned by $2\pi i \operatorname{diag}(r_0,\ldots,r_n)$. Let $a \in \mathbf{k}^*$ be a basis element of norm 1. By 2.7 if $x = (x_0:\ldots:x_n) \in X$ then

$$\mu(x) = \Big(\sum_{0\leq j\leq n} r_j |x_j|^2\Big)\Big(\sum_{0\leq j\leq n} |x_j|^2\Big)^{-1} a.$$

Now by theorem 6.18 $x \in X^{min}$ if and only if $0 \in \mu(\overline{Gx})$. The map from $G = \mathbf{C}^*$ to X given by

$$z \to (z^{r_0} x_0:\ldots:z^{r_n} x_n)$$

extends uniquely to a map $\theta\colon \mathbf{P}_1 \to X$ with

$$\theta(0) = (y_0:\ldots:y_n)$$

where $y_j = x_j$ if $r_j = \min\{r_i \mid x_i \neq 0\}$ and $y_j = 0$ otherwise, and

$$\theta(\infty) = (y_0':\ldots:y_n')$$

where $y_j' = x_j$ if $r_j = \max\{r_i \mid x_i \neq 0\}$ and $y_j' = 0$ otherwise. Then $\overline{Gx}$ is the image of $\mathbf{P}_1$ under θ and

$$\mu(\theta(0)) = \min\{r_j \mid x_j \neq 0\}$$

while

$$\mu(\theta(\infty)) = \max\{r_j \mid x_j \neq 0\}.$$

On the other hand $0 \in \mu(\overline{Gx})$ if and only if either $r_j = 0$ whenever $x_j \neq 0$ or

$$\min\{r_j \mid x_j \neq 0\} < 0 < \max\{r_j \mid x_j \neq 0\}.$$

This is because if

$$\min\{r_j \mid x_j \neq 0\} < 0 < \max\{r_j \mid x_j \neq 0\}$$

then

$$\sum_j r_j |x_j|^2 |z|^{2r_j}$$

tends to ∞ as $|z| \to \infty$ and tends to $-\infty$ as $|z| \to 0$. It follows that $0 \in \mu(\overline{Gx})$ if and only if

$$\min\{r_j \mid x_j \neq 0\} \leq 0 \leq \max\{r_j \mid x_j \neq 0\},$$

and thus by 8.5 $x \in X^{min}$ if and only if $x \in X^{ss}$.

In order to deduce from this lemma that $X^{ss} = X^{min}$ in the general case, we investigate next the relationship between the minimum strata associated to the action of the whole maximal compact subgroup K and of its closed real one-parameter subgroups $\lambda: S^1 \to K$.

8.7. <u>Definition.</u> A complex 1-PS (one-parameter subgroup) $\lambda: \mathbf{C}^* \to G$ of G will be called <u>compatible with</u> K if it is the complexification of a closed real 1-PS $\lambda: S^1 \to K$ of K. When λ is compatible with K let μ_λ be the composition of μ with $\lambda^*: \mathbf{k}^* \to \mathbf{R}$. Then μ_λ is a moment map for the action of S^1 on X via λ.

8.8. <u>Lemma. If</u> $x \in X$ <u>then</u> $0 \in \mu(\overline{Gx})$ <u>if and only if</u> $0 \in \mu_\lambda(\overline{\lambda(\mathbf{C}^*)x})$ <u>for every</u> 1-PS $\lambda: \mathbf{C}^* \to G$ <u>compatible with</u> K. <u>Equivalently the minimum stratum</u> X^{min} <u>is the intersection of the minimum strata</u> X_λ^{min} <u>associated to</u>

the action on X of all the 1-PSs λ compatible with K.

8.9. Remark. The proof of this lemma is valid when X is any Kähler manifold and G is the complexification of a compact subgroup K which preserves the Kähler structure on X. We are going to see that when X is a complex projective variety then $X^{min} = X^{ss}$. Therefore this result can be regarded as a generalisation to Kähler manifolds of the fundamental fact of geometric invariant theory, which says that a point is semistable for the action of a group if and only if it is semistable for the action of every 1-PS.

Proof. First note that the proof given at 6.19 shows that $X^{min} \subseteq X_{\lambda}^{min}$ for every 1-PS λ of G compatible with K.

Now suppose x does not lie in X^{min}. Then there exists a nonzero $\beta \in \mathbf{B}$ such that $x \in S_\beta$. By 6.18 $S_\beta = GY_\beta^{min}$ and this is the same as KY_β^{min} since Y_β^{min} is invariant under the parabolic subgroup P_β and $G = KP_\beta$. Therefore $kx \in Y_\beta^{min}$ for some $k \in K$.

Now $\mu(X)$ is compact in $\mathbf{k}^* \cong \mathbf{k}$ and the rational points are dense in $\mathbf{t}$. Therefore there exists $\delta > 0$ and a rational point $\gamma \in \mathbf{t}$ such that

$$\{\xi \in \mu(X) \mid \xi.\beta \geq \|\beta\|^2\} \subseteq \{\gamma \in \mathbf{k} \mid \xi.\gamma \geq \delta\}.$$

By replacing γ by $m\gamma$ for a suitable integer $m > 0$ we may assume that γ is a lattice point of $\mathbf{t}$ and hence corresponds to a complex 1-PS of T_c compatible with T. Since $kx \in Y_\beta^{min}$, by 6.11 we have

$$\mu(\overline{\gamma(\mathbf{C}^*)kx}) \subseteq \{\xi \in \mu(X) \mid \xi.\beta \geq \|\beta\|^2\} \subseteq \{\xi \in \mathbf{k} \mid \xi.\gamma \geq \delta\}.$$

In particular $\mu_\gamma(\overline{\gamma(\mathbf{C}^*)kx})$, which is the projection along γ of $\mu(\overline{\gamma(\mathbf{C}^*)kx})$, does not contain 0. Let $\lambda = \mathrm{Ad}(k)\gamma$. Then λ is a 1-PS of G compatible with K such that $0 \notin \mu_\lambda(\overline{\lambda(\mathbf{C}^*)x})$ and hence $x \notin X_\lambda^{min}$. Therefore $\bigcap_\lambda X_\lambda^{min} \subseteq X^{min}$, and the proof is complete.

Any 1-PS $\lambda: \mathbf{C}^* \to G$ has a conjugate $\mathrm{Ad}(g)\lambda = g\lambda g^{-1}: \mathbf{C}^* \to G$ which is compatible with K. Therefore from 8.4, 8.6, 8.8 and the fact that X^{min} is G-invariant we can deduce the following

8.10. Theorem. Let $X \subseteq \mathbf{P}_n$ be a nonsingular complex projective variety and let G be a complex reductive algebraic group acting on X via a homomorphism $\phi: G \to GL(n+1)$. Suppose that G has a maximal compact subgroup K such that $\phi(K) \subseteq U(n+1)$. Then the set X^{ss} of semistable points of X coincides with the minimum Morse stratum X^{min} of the function $|\mu|^2$ on X, where $\mu: X \to \mathbf{k}^*$ is the moment map defined at 2.7 and $\| \ \|$ is the norm associated to any K-invariant inner product on $\mathbf{k}$.

Suppose now that the stabiliser in G of every semistable point is finite. Then by remark 8.3 we have $X^s = X^{ss}$. But we know that there is a morphism $\psi: X^{ss} \to M$ from X^{ss} to the projective quotient M such that each fibre which meets X^s is a single orbit under the action of G (see [M] theorem 1.10). Therefore ψ induces a continuous bijection $\tilde{\psi}: X^{ss}/G \to M$.

We saw in §7 that X^{ss}/G is a compact Hausdorff space, and so is the projective variety M. Hence $\tilde{\psi}$ is a homeomorphism.

Thus we obtain formulae for the rational cohomology of the quotient variety M. Before stating these formulae in a theorem, let us review the definitions of the terms involved and interpret them in the case of a linear reductive action on a projective variety.

First recall from 3.5 that the moment map μ_T for the action of the compact maximal torus T on X is given by

$$\mu_T(x) = (\sum_j |x_j|^2)^{-1} \sum_j |x_j|^2 \alpha_j$$

where $\alpha_0,\ldots,\alpha_n$ are the weights of the action.

Choose an inner product which is invariant under the Weyl group action on the Lie algebra $\mathbf{t}$ of T and use it to identify $\mathbf{t}^*$ with $\mathbf{t}$. Then a minimal combination of weights is by definition the closest point to 0 of the convex hull of some nonempty subset of $\{\alpha_0,\ldots,\alpha_n\}$. The indexing set $\mathbf{B}$ consists of all minimal weight combinations lying in the positive Weyl chamber $\mathbf{t}_+$.

Note that if we assume the inner product to be rational (i.e. to take rational values on lattice points) then <u>each</u> $\beta \in \mathbf{B}$ <u>is a rational point of</u> $\mathbf{t}$. Thus each subgroup $\exp\mathbf{R}\beta$ of T is closed and hence the subtorus T_β of T generated by β is one-dimensional.

We saw in 3.11 that for each $\beta \in \mathbf{B}$ the submanifold Z_β of X is the intersection of X with the linear subspace

$$\{x \in \mathbf{P}_n \mid x_j = 0 \text{ unless } \alpha_j.\beta = \|\beta\|^2\}$$

of $\mathbf{P}_n$. Recall that Z_β^{min} was defined as the set of points in Z_β whose paths of steepest descent for the function $|\mu-\beta|^2$ on Z_β have limit points in $Z_\beta \cap \mu^{-1}(\beta)$. Let $\mathrm{Stab}\beta$ be the stabiliser of β under the adjoint action of G and let $\mathrm{Stab}_K\beta$ be its intersection with K. By 4.9 $\mu - \beta$ is a moment map for the action of $\mathrm{Stab}_K\beta$ on Z_β.

8.11. In order to interpret the inductive formula of 5.10 we want to define a subset Z_β^{ss} of Z_β somehow in terms of semistability so that Z_β^{ss} will coincide with Z_β^{min}. There are at least two alternative ways to do this. One way is to let G_β be the complexification of the connected closed subgroup of $\mathrm{Stab}_K\beta$ whose Lie algebra is the orthogonal complement to β, and to let Z_β^{ss} be the set of points of Z_β which are semistable for the linear action of G_β on Z_β defined by the homomorphism ϕ. Then $Z_\beta^{ss} = Z_\beta^{min}$ by theorem 8.10 because the projection onto the Lie algebra of $K \cap G_\beta$ of μ restricted to Z_β is $\mu - \beta$. Another way is to note that since β is a rational point of the centre of $\mathrm{stab}\beta$ there is a character $\chi\colon \mathrm{Stab}\beta \to \mathbf{C}^*$ whose derivative is a positive integer multiple $r\beta$ of β. One can define Z_β^{ss} to be the set of semistable points of Z_β under the action of $\mathrm{Stab}\beta$, where the action is linearised with respect to the rth tensor power of the hyperplane bundle by the product of $\phi^{\otimes r}$ with the inverse of the character χ. The corresponding moment map Z_β is then $r\mu - r\beta$ so that again $Z_\beta^{ss} = Z_\beta^{min}$. However the details are unimportant.

$Z_{\beta,m}$ can be reinterpreted as the union of those components of Z_β which are contained in components of Y_β of real codimension m. Finally β-sequences $\underline{\beta} = (\beta_1,\ldots,\beta_q)$ and the corresponding linear sections $Z_{\underline{\beta}}$ and $Z_{\underline{\beta},m}$ of X and subgroups $\mathrm{Stab}\underline{\beta}$ can be defined as in §5.

The theorem for which we have been aiming all along can now be stated.

8.12. Theorem. Let $X \subseteq \mathbf{P}_n$ be a complex projective variety acted on linearly by a connected complex reductive algebraic group G. The equivariant Poincaré series for X^{ss} is given by the inductive formula

$$P_t^G(X^{ss}) = P_t(X)P_t(BG) - \sum_{\beta,m} t^{d(\beta,m)} P_t^{\mathrm{Stab}\beta}(Z_{\beta,m}^{ss})$$

where the sum is over nonzero $\beta \in \mathbf{B}$ and integers $0 \leq m \leq \dim X$. $\mathrm{Stab}\beta$ is a reductive subgroup of G acting on $Z_{\beta,m}$ which is a smooth subvariety of X for each β and m, and

$$d(\beta,m) = m - \dim G + \dim \mathrm{Stab}\beta.$$

Suppose that the stabiliser of every semistable point in X is finite, so that the projective quotient variety M associated to the action in geometric invariant theory is topologically the quotient X^{ss}/G. Then the rational cohomology of M is isomorphic to the G-equivariant rational cohomology of X^{ss}. It is given by the explicit formula

$$P_t(M) = P_t(X)P_t(BG) + \sum_{\underline{\beta},m} (-1)^q t^{d(\underline{\beta},m)} P_t(Z_{\underline{\beta},m})P_t(B\ \mathrm{Stab}\underline{\beta})$$

where the sum is over β*-sequences* $\underline{\beta} = (\beta_1,\ldots,\beta_q)$ *and integers* $0 \leq m \leq \dim X$. *Each* $Z_{\underline{\beta},m}$ *is a smooth subvariety of* X *acted on by a reductive subgroup* $\mathrm{Stab}\underline{\beta}$ *of* G, *and*

$$d(\beta,m) = m - \dim G + \dim \mathrm{Stab}\underline{\beta}.$$

Proof. This follows immediately from 5.10, 5.16, 6.20, 8.10 and the remarks of the last few paragraphs.

8.13. *Remark.* Note that equivariant cohomology must be used in the inductive formula because the conditon of finite isotropy groups may not be satisfied for all the subgroups $\mathrm{Stab}\beta$ acting on the subvarieties $Z_{\beta,m}$.

8.14. *Remark.* When the stabiliser of every semistable point is finite then the geometric invariant theory quotient $\mu = X^{ss}/G$ is homeomorphic to the symplectic quotient $\mu^{-1}(0)/K$ by theorem 7.5. In fact we can show that M *is homeomorphic to* $\mu^{-1}(0)/K$ without any assumption on stabilisers as follows (cf. [Ne] theorem 2.4).

The inclusions $\mu^{-1}(0) \to X^{min} = X^{ss}$ together with the surjective G-invariant morphism ψ: $X^{ss} \to M$ induce a continuous map

$$h: \mu^{-1}(0)/K \to M.$$

By the proofs of [M] theorem 1.10 and amplification 1.3 two points x and y in X^{ss} are identified by ψ if and only if the closures in X^{ss} of Gx and Gy meet each other. But by remark 7.8 $G\mu^{-1}(0)$ consists of those $x \in X^{ss}$ such that Gx is closed in X^{ss}, so the map $h: \mu^{-1}(0)/K \to M$ is injective. Moreover if $x \in X^{ss} = X^{min}$ then the closure of the path of steepest descent for the function $\|\mu\|^2$ from x contains a point of $\mu^{-1}(0)$, and by 6.7 this path is contained in the orbit Gx. Thus h is surjective. It follows that h is a bijection from a compact space to a Hausdorff space, and hence is a homeomorphism.

§9. Some remarks on non-compact manifolds

So far we have considered only compact symplectic manifolds and projective varieties. Now suppose X is any symplectic manifold acted on by a compact group K such that a moment map $\mu: X \to k^*$ exists. Then one can obtain almost the same results as for compact manifolds subject only to the condition that

9.1. for some metric on X, every path of steepest descent under the function $f = \|\mu\|^2$ is contained in some compact subset of X.

One simply checks that all the arguments used in §§3,4,5 and the appendix are still valid with trivial modifications. The only result which fails is theorem 5.8. This says that the rational equivariant cohomology of the total space X is the tensor product of its ordinary rational cohomology with that of the classifying space of the group K; i.e. that

$$P_t^K(X) = P_t(X)P_t(BK).$$

Thus in the formulae obtained for the equivariant rational cohomology of $\mu^{-1}(0)$ (see 5.10 and 5.16) one must now always use the equivariant Poincaré series $P_t^K(X)$ rather than the product $P_t(X)P_t(BK)$. Otherwise the formulae are correct and in good cases give the Betti numbers of the symplectic quotient $\mu^{-1}(0)/K$.

9.2. Example: cotangent bundles. The examples which motivated the definition of symplectic manifolds and moment maps were phase spaces and conserved quantities such as angular momentum.

The cotangent bundle T^*M of any manifold M has a natural symplectic structure given by

$$\omega = \sum_i dp_i \wedge dq_i$$

where $(q_1,\ldots,q_n)$ are local coordinates on M and $(p_1,\ldots,p_n)$ are the induced coordinates on the cotangent space at $(q_1,\ldots,q_n)$. Any action of a compact group K on M induces an action of K on T^*M which preserves this symplectic structure. Moreover it is not hard to check that there is a moment map

$$\mu: T^*M \to \mathbf{k}^*$$

for this action defined as follows. If $m \in M$ and $\xi \in T_m^*M$ then

9.3 $$\mu(\xi).a = \xi.a_m$$

for all $a \in \mathbf{k}$, where . on the left hand side denotes the natural pairing between $\mathbf{k}^*$ and $\mathbf{k}$ and on the right denotes the natural pairing between T_m^*M and T_mM. So a general moment map is of the form $\mu + c$ where c lies in the centre of $\mathbf{k}^*$ (see §2).

The condition 9.1 holds for each of the moment maps on T^*M provided that M is compact. To see this one fixes a metric on M and uses it to induce a Riemannian metric on T^*M. It can then be shown that the path of steepest descent for the function $f = \|\mu + c\|^2$ from any point $\xi \in T^*M$

consists of cotangent vectors whose norm is bounded by some number depending only on ξ.

The function $f = |\mu|^2$ where μ is given by 9.3 is not an interesting Morse function because the only critical points are the points in $\mu^{-1}(0)$. The reason for this is that by lemma 3.1 if $\xi \in T_m^* M$ is critical for f then the vector field induced by $\mu(\xi)$ on T^*M vanishes at ξ. Thus in particular $\mu(\xi)_m = 0$, so if we put $a = \mu(\xi)$ in 9.3 we obtain $\|\mu(\xi)\|^2 = 0$. However if K is not semisimple then it is often possible to choose c in the centre of **k** such that the norm-square of the moment map $\mu + c$ has non-minimal critical points.

For example, consider the action of the circle S^1 on T^*S^2 induced by rotation of the sphere S^2 about some axis. Let c be an element of norm 1 in the Lie algebra of S^1 and let $f = \|\mu + c\|^2$. Then from 9.3 we have

$$f(\xi) = (c_m . \xi + 1)^2$$

for any $m \in S^2$ and $\xi \in T_m^* S^2$. So $f(\xi) = 0$ if and only if $\xi . c_m = -1$, which means that the minimum set for f is homeomorphic to a line bundle over the sphere less two points and hence is homotopically equivalent to S^1. Since the circle action on this is free the equivariant cohomology of the minimum set is trivial.

By lemma 3.1 the other critical points ξ for f are those fixed by S^1. These are the two points of S^2 fixed by the rotation. The index of the Hessian at each of these is 2. Thus we obtain

$$P_t^{S^1}(S^2) = P_t^{S^1}(T^*S^2) = 1 + 2t^2(1-t^2)^{-1} = (1+t^2)(1-t^2)^{-1}$$
$$= P_t(S^2)P_t(BS^1)$$

as one expects from proposition 5.8 since S^2 has a symplectic structure preserved by the action of S^1.

As a second example, consider the linear action of the torus

$$T = \{ \begin{bmatrix} e^{i\theta} & 0 \\ 0 & e^{i\phi} \end{bmatrix} \mid \theta,\phi \in \mathbf{R}\}$$

on the unit sphere $S^3 \subseteq \mathbf{C}^2$. By 9.3 if $m \in S^3$ and $\xi \in T_m^*S^3$ then

$$\mu(\xi) = (a_m.\xi)a + (b_m.\xi)b$$

where $a = \begin{bmatrix} i & 0 \\ 0 & 0 \end{bmatrix}$ and $b = \begin{bmatrix} 0 & 0 \\ 0 & i \end{bmatrix}$. Consider the function

$$f = \|\mu + a + b\|^2$$

on T^*S^3. Any $\xi \in T_m^*S^3$ satisfies $f(\xi) = 0$ if $\mu(\xi) = -a - b$, i.e. if

$$a_m.\xi = -1 = b_m.\xi .$$

If $a_m = 0$ or $b_m = 0$ these equations for ξ have no solution, and otherwise they define an affine line in $T_m^*S^3$. So the minimum set $f^{-1}(0)$ is acted on freely by T and its equivariant cohomology is isomorphic to the cohomology of the quotient by T of S^3 with two circles removed. This quotient is an open interval, so its cohomology is trivial.

From lemma 3.1 we see that if $\xi \in T_m^*S^3$ is a non-minimal critical point for f then either ξ is fixed by a and $\mu(\xi) + a + b$ is a scalar multiple of a, or ξ is fixed by b and $\mu(\xi) + a + b$ is a scalar multiple of b. In the first case $\xi.b = -1$ and $\xi \in T^*S^1$ where S^1 is the circle fixed by a, and the second case is similar. So the non-minimal critical points form two circles in T^*S^3, each of which is fixed by one copy of S^1 in the torus T and is acted on freely by the other. The index of the function f along each of these circles is 2. Thus we obtain

$$P_t^T(S^3) = 1 + 2t^2(1 - t^2)^{-1} .$$

Note that this is not equal to $P_t(S^3)P_t(BT)$; this does not contradict proposition 5.8 because S^3 is not a symplectic manifold.

9.4. Example: quasi-projective varieties. Other obvious examples of non-compact symplectic manifolds are nonsingular quasi-projective complex varieties.

Suppose G is a complex reductive group with maximal compact subgroup K, acting linearly on a nonsingular locally closed subvariety X of some complex projective space $\mathbf{P}_n$. Suppose also that the stabiliser of every semistable point is finite. If condition 9.1 is satisfied then we obtain formulae for the Betti numbers of the symplectic quotient $\mu^{-1}(0)/K$ which is homeomorphic to the quotient variety produced by invariant theory (cf. §8). There is also a more algebraic condition for these formulae to exist which is an alternative to 9.1. It is described as follows.

When X is a closed subvariety of $\mathbf{P}_n$ acted on linearly by G then the stratification of X induced by the action is just the intersection with X of the stratification $\{S_\beta \mid \beta \in \mathbf{B}\}$ induced on $\mathbf{P}_n$. If X is quasi-projective we can still define a stratification of X with strata $\{X \cap S_\beta \mid \beta \in \mathbf{B}\}$. Moreover by 6.18 and 8.10 we have

$$S_\beta \cong G \times_{P_\beta} Y_\beta^{ss}$$

for each $\beta \in \mathbf{B}$ where Y_β^{ss} is a nonsingular locally-closed subvariety of $\mathbf{P}_n$ and P_β is a parabolic subgroup of G. Since X is invariant under G this implies that

$$X \cap S_\beta \cong G \times_{P_\beta} (X \cap Y_\beta^{ss}).$$

There is also a retraction $p_\beta : Y_\beta^{ss} \to Z_\beta^{ss}$ of Y_β^{ss} onto the semistable points of a linear subvariety Z_β of $\mathbf{P}_n$ under the action of a subgroup of G. Provided that

9.5. $p_\beta(x) \in X$ <u>whenever</u> $x \in X \cap Y_\beta^{ss}$ <u>for each</u> $\beta \in \mathbf{B}$,

one can check that each p_β induces a retraction of $X \cap Y_\beta^{ss}$ onto $X \cap Z_\beta^{ss}$ and that all the results of §8 hold for X.

One can use quasi-projective varieties satisfying this condition to rederive Atiyah and Bott's formulae for the cohomology of moduli spaces of vector bundles over Riemann surfaces (see [Ki3]). For this one considers spaces of holomorphic maps from Riemann surfaces to Grassmannians. These can be embedded as quasi-projective subvarieties of products of Grassmannians.

The results of Part I also apply to reductive group actions on singular varieties satisfying appropriate conditions (see the work of Carrell and Goresky on $\mathbf{C}^*$ actions [C & G]).

§10. Appendix: Morse theory extended to minimally degenerate functions

Given any nondegenerate Morse function with isolated critical points on a compact manifold, one has the well-known Morse inequalities which relate the Betti numbers of the manifold to the numbers of critical points of each index. Bott has shown that this classical Morse theory extends to a more general class of Morse functions (see [Bo]). The functions which are nondegenerate in the sense of Bott are those whose critical sets are disjoint unions of submanifolds along each of which the Hessian is nondegenerate in normal directions. The associated Morse inequalities relate the Betti numbers of the manifold to the Betti numbers and indices of the critical submanifolds. The purpose of this appendix is to show that Morse theory can be extended to cover an even larger class of functions.

10.1. Definition. A smooth function $f\colon X \to \mathbf{R}$ on a compact manifold X is called minimally degenerate if the following conditions hold.

(a) The set of critical points for f on X is a finite union of disjoint closed subsets $\{C \in \mathbf{C}\}$ on each of which f takes a constant value $f(C)$. The subsets C are called critical subsets of f. If the critical set of f is reasonably well-behaved we can take the subsets $\{C \in \mathbf{C}\}$ to be its connected components.

(b) For every $C \in \mathbf{C}$ there is a locally closed submanifold Σ_C containing C and with orientable normal bundle in X such that

(i) C is the subset of Σ_C on which f takes its minimum value,

and

(ii) at every point $x \in C$ the tangent space $T_x\Sigma_C$ is maximal among all subspaces of T_xX on which the Hessian $H_x(f)$ is positive semi-definite.

A submanifold satisfying these properties is called a minimising manifold for f along C.

Thus minimal degeneracy means that critical sets can be as degenerate as a minimum but no worse.

The purpose behind this definition is to find a condition on f more general than nondegeneracy which ensures that for some choice of metric f induces a Morse stratification whose strata are all smooth. This appendix shows that minimal degeneracy is such a condition. Conversely if f is any function which induces a smooth Morse stratification then the strata themselves are minimising manifolds provided that the Hessian at every critical point is definite in directions normal to the stratum which contains it.

We do not demand that the minimising manifolds be connected. However, this extra condition is always satisfied if we replace each critical subset C by its intersections with the connected components of Σ_C. Hence we can assume that the index of the Hessian of f takes a constant value $\lambda(C)$ along any $C \in \mathcal{C}$, because by 10.1 (ii) it coincides with the codimension of the submanifold Σ_C. We shall call $\lambda(C)$ the index of f along C.

Any function f which is nondegenerate in the sense of Bott is minimally degenerate. For by definition the set of critical points of f is the disjoint union of connected submanifolds of X and these can be taken as the critical subsets of f. If we fix a metric on X then the Hessian of f induces a self-adjoint endomorphism of the normal bundle N_C along each critical submanifold C. Because f is nondegenerate N_C splits as a sum $N_C^+ \oplus N_C^-$ where the Hessian is positive definite on N_C^+ and negative definite on N_C^-. It is easy to check that locally the image of N_C^+ under the exponential map induced by the metric is a minimising manifold for f along C.

We wish to show that any minimally degenerate Morse function on X induces Morse inequalities in cohomology, and also in equivariant cohomology if X is acted on by a compact group K which preserves the function. These inequalities are most easily expressed using Poincaré polynomials

$$P_t(X) = \sum_{j \geq 0} t^j \dim H^j(X)$$

and equivariant Poincaré polynomials

$$P_t^K(X) = \sum_{j \geq 0} t^j \dim H_K^j(X) .$$

Our aim is to prove the following theorem.

10.2. <u>Theorem.</u> <u>Let</u> $f: X \to \mathbf{R}$ <u>be a minimally degenerate Morse function with critical subsets</u> $\{C \in \mathbf{C}\}$ <u>on a compact manifold</u> X. <u>Then the Betti numbers of</u> X <u>satisfy Morse inequalities which can be expressed in the form</u>

$$\sum_{C \in \mathbf{C}} t^{\lambda(C)} P_t(C) - P_t(X) = (1 + t)R(t)$$

where $\lambda(C)$ is the index of f along C and

$$R(t) \geq 0$$

in the sense that all its coefficients are non-negative. If a compact group K acts on X preserving f and the minimising manifolds, then X also satisfies equivariant Morse inequalities of the same form.

When f is nondegenerate one method of obtaining the Morse inequalities is to use a metric to define a smooth stratification $\{S_C | C \in \mathbf{C}\}$ of X. This is perhaps not the easiest approach, but we shall follow it here because the stratification of the particular function relevant to us is interesting in its own right. A point of X lies in a stratum S_C if its trajectory under the gradient field $-\text{grad } f$ converges to a point of the corresponding critical subset C. For a general function f such a trajectory may not converge to a single point. However the limit set of the trajectory is always a connected nonempty set of critical points for f (see 2.10). Therefore if f is minimally degenerate then any such limit set is contained in a unique critical subset. So we make the following definition.

10.3. Definition. Suppose $f: X \to \mathbf{R}$ is a minimally degenerate Morse function with critical subsets $\{C \in \mathbf{C}\}$ and suppose X is given a fixed Riemannian metric. Then for each $C \in \mathbf{C}$ let S_C be the subset of X

consisting of all points $x \in X$ such that the limit set $\omega(x)$ of the trajectory of $-\text{grad } f$ from x is contained in C.

X is the disjoint union of the subsets $\{S_C | C \in \mathbf{C}\}$. We shall see that if the metric is chosen appropriately they form a smooth stratification of X such that each stratum S_C coincides near C with the minimising manifold Σ_C. The condition which the metric must satisfy is that the gradient field grad f should be tangential to each minimising manifold Σ_C. We shall show that such a metric exists, and then prove the following theorem.

10.4. Theorem. Let $f: X \to \mathbf{R}$ be a minimally degenerate Morse function with critical subsets $\{C \in \mathbf{C}\}$ on a compact Riemannian manifold. Suppose that the gradient flow of f is tangential to each of the minimising manifolds $\{\Sigma_C | C \in \mathbf{C}\}$. Then the subsets $\{S_C | C \in \mathbf{C}\}$ defined at 10.3 form a smooth stratification of X, called the Morse stratification of the function f on X. For each $C \in \mathbf{C}$ the stratum S_C coincides with the minimising manifold Σ_C in some neighbourhood of C. Moreover each inclusion $C \to S_C$ is an equivalence of Čech cohomology. If there is a compact group K acting on X such that the function f, the minimising manifolds and the metric are invariant under K then these inclusions are also equivalences of equivariant cohomology.

In order to be able to apply this result to any minimally degenerate function we need the following lemma.

10.5. Lemma. Let $f: X \to \mathbf{R}$ be a minimally degenerate function on X. Then there exists a metric on X such that near each $C \in \mathbf{C}$ the gradient flow of f is tangential to the minimising manifold Σ_C. If f and the minimising manifolds are invariant under the action of a compact group K then the metric may be taken to be K-invariant.

Proof. A standard argument using partitions of unity shows that it is enough to find such metrics locally. The only point to note is that one should work with dual metrics because $\text{grad}_\rho f$ is linear in ρ^* but not in ρ.

Suppose x is any point of a critical subset C. Condition (ii) of 10.1 implies that there is a complement to $T_x\Sigma_C$ in T_xX on which the Hessian $H_x(f)$ is negative definite. It follows from the Morse lemma (lemma 2.2 of [Mi]) that there exist local coordinates $(x_1,\ldots,x_n)$ around x such that the minimising manifold Σ_C is given locally by

$$0 = x_{d+1} = x_{d+2} = \cdots = x_n ,$$

and such that

$$f(x_1,\ldots,x_n) = f(x_1,\ldots,x_d) - (x_{d+1})^2 - \cdots - (x_n)^2.$$

(To prove this regard $x_1,\ldots,x_d$ as parameters and apply the Morse lemma to $x_{d+1},\ldots,x_n$). Then the gradient flow of f with respect to the standard metric on $\mathbf{R}^n$ is tangential to Σ_C.

Finally a K-invariant metric is obtained by averaging the dual metric over K.

Because of this lemma, theorem 10.2 can be deduced from theorem 10.4 by the standard argument using Thom-Gysin sequences (cf. §2). The rest of this appendix is devoted to the proof of theorem 10.4.

The most difficult part of the proof of this theorem will be to show that for each $C \in \mathbf{C}$ the stratum S_C coincides with the given submanifold Σ_C in some neighbourhood of C. Once we know that S_C is smooth near C it will follow easily that S_C is smooth everywhere, and the cohomology equivalences are not hard to prove.

First we shall show that the subsets $\{S_C | C \in \mathbf{C}\}$ form a stratification of X in the sense of 2.11. It suffices to prove the following lemma, which depends on the assumption 10.1(a) but not on the existence of minimising manifolds.

10.7. Lemma. For each $C \in \mathbf{C}$,

$$\overline{S}_C \subseteq S_C \cup \bigcup_{f(C')>f(C)} S_{C'} \,.$$

Proof. If a point x lies in S_C for some $C \in \mathbf{C}$ then by definition its path of steepest descent for f has a limit point in C, and hence

$$f(x) \geq f(C)$$

since f decreases along this path. Moreover

$$f(x) > f(C)$$

unless $x \in C$.

If x lies in the closure $\overline{S}_C$ of S_C then so does every point of its path of steepest descent. Hence the closure of this path is contained in $\overline{S}_C$. It follows that $x \in S_{C'}$ for some C' with $f(C') \geq f(C)$. So if $x \in \overline{S}_C$ and x is not critical for f then $f(x) > f(C)$.

Since the subsets $\{C \in \mathbf{C}\}$ are compact there are open sets $\{U_C \mid C \in \mathbf{C}\}$ whose closures are disjoint such that $U_C \supseteq C$ for each $C \in \mathbf{C}$. If x lies in the boundary ∂U_C of some U_C then x is not critical for f. Hence if $x \in \partial U_C \cap \overline{S}_C$ then $f(x) > f(C)$. Since each $\partial U_C \cap \overline{S}_C$ is compact it follows that there is some $\delta > 0$ such that if $C \in \mathbf{C}$ and $x \in \partial U_C \cap \overline{S}_C$ then $f(x) \geq f(C) + \delta$.

Now suppose that C and C' are distinct and that there is some $x \in S_{C'} \cap \overline{S}_C$. Let $\{x_t \mid t \geq 0\}$ be the path of steepest descent for f with $x_0 = x$. Then the limit points of $\{x_t \mid t \geq 0\}$ as $t \to \infty$ are contained in C'. So there exists $T \geq 0$ such that $x_T \in U_{C'}$ and $f(x_T) < f(C') + \delta$. But this implies that there is a neighbourhood V of x in X such that $y_T \in U_{C'}$ and $f(y_T) < f(C') + \delta$ whenever $y \in V$.

Since $x \in \overline{S}_C$ there is some $y \in V \cap S_C$. Then $y_T \in U_{C'}$ but the limit points as $t \to \infty$ of $\{y_t \mid t \geq 0\}$ are contained in C. Since by assumption $\overline{U}_C \cap \overline{U}_{C'} = \emptyset$ there must exist some $t > T$ such that $y_t \in \partial U_C \cap S_C$. This implies that $f(y_t) \geq f(C) + \delta$ by the choice of δ. But f decreases along the path $\{y_t \mid t \geq 0\}$ and $f(y_T) < f(C') + \delta$ since $y \in V$. Therefore

$$f(C') + \delta > f(y_T) \geq f(y_t) \geq f(C) + \delta,$$

so that

$$f(C') > f(C).$$

This shows that if $S_{C'} \cap \overline{S}_C$ is nonempty then $f(C) < f(C')$. Since X is the disjoint union of the subsets $\{S_C \mid C \in \mathbf{C}\}$ the result follows.

Now we shall begin the proof that each stratum S_C coincides near C with the corresponding minimising manifold Σ_C.

10.8 Lemma. For each $C \in \mathbf{C}$ the intersection of the minimising manifold Σ_C with a sufficiently small neighbourhood of C is contained in the Morse stratum S_C.

Proof. As in the proof of 10.7 choose open subsets $\{U_C | C \in \mathbf{C}\}$ of X whose closures are disjoint and such that $U_C \supseteq C$ for each $C \in \mathbf{C}$. Since each Σ_C is a submanifold of some neighbourhood of C, if U_C is taken small enough then $\Sigma_C \cap \bar{U}_C$ is closed for each $C \in \mathbf{C}$.

If $C \in \mathbf{C}$ then by the definition of a minimising manifold C is the subset of Σ_C on which f takes its minimum value. Hence if $x \in \Sigma_C \cap \partial U_C$ then $f(x) > f(C)$, and so as $\Sigma_C \cap \partial U_C$ is compact there exists $\gamma > 0$ such that

$$f(x) \geq f(C) + \gamma$$

whenever $C \in \mathbf{C}$ and $x \in \Sigma_C \cap \partial U_C$. Then for every $C \in \mathbf{C}$ the subset

$$V_C = U_C \cap \{x \in X | f(x) < f(C) + \gamma\}$$

is an open neighbourhood of C in X.

Suppose x lies in the intersection of this neighbourhood V_C with Σ_C. Then as grad f is tangential to Σ_C and Σ_C is closed in $\bar{U}_C$ the path $\{x_t | t \geq 0\}$ of steepest descent for f from x stays in Σ_C as long as it remains in U_C. Hence if the path leaves U_C there exists $t > 0$ such that $x_t \in \partial U_C \cap \Sigma_C$. This implies that

$$f(x) \geq f(x_t) \geq f(C) + \gamma,$$

which contradicts the assumption that $x \in V_C$. So the path remains in U_C for all time. Since the only critical points for f in $\overline{U}_C$ are contained in C it follows that the limit points of the path lie in C and so $x \in S_C$.

10.9. Remark. Note that the same argument shows that given any neighborhood U_C of C in X there exists a smaller neighbourhood V_C such that if $x \in V_C \cap S_C$ then the entire path of steepest descent for f from x is contained in U_C.

In order to prove the converse to the last lemma we need to investigate the differential equation which defines the gradient flow of f in local coordinates near any critical point x. We shall rely on the standard local results to be found in [H].

Recall that if $x \in X$ is a critical point for f then the Hessian $H_x(f)$ of f at x is a symmetric bilinear form on the tangent space T_xX given in local coordinates by the matrix of second partial derivatives of f. The Riemannian metric provides an inner product on T_xX so that H_x can be identified with a self-adjoint linear endomorphism of T_xX. Then all the eigenvalues of $H_x(f)$ are real and T_xX splits as the direct sum of the eigenspaces of $H_x(f)$.

The assumption that the gradient field of f is tangential to Σ_C implies that for each $x \in C$ the subspace $T_x\Sigma_C$ of T_xX is invariant under $H_x(f)$

regarded as a self-adjoint endomorphism of T_xX. Hence so is its orthogonal complement $T_x\Sigma_C^{\perp}$. By the definition of a minimising manifold the eigenvalues of $H_x(f)$ restricted to $T_x\Sigma_C$ are all non-negative, while those of $H_x(f)$ restricted to $T_x\Sigma_C^{\perp}$ are all strictly negative.

Now fix $C \in \mathbf{C}$ and a point $x \in C$. Let d be the dimension of Σ_C. Then we can find local coordinates $(x_1,\ldots,x_n)$ in a neighbourhood W_x of x such that

10.10. (i) x is the origin in these coordinates and the submanifold Σ_C is given by $x_{d+1} = x_{d+2} = \ldots = x_n = 0$;
(ii) the Riemannian metric at x is the standard inner product on $\mathbf{R}^n$;
and
(iii) the Hessian $H_x(f)$ is represented by a diagonal matrix

$$H_x(f) = \operatorname{diag}(\lambda_1,\ldots,\lambda_n)$$

where

$$\lambda_1,\ldots,\lambda_d \geq 0$$

and

$$\lambda_{d+1},\ldots,\lambda_n < 0.$$

Let P be the diagonal matrix $\operatorname{diag}(-\lambda_1,\ldots,-\lambda_d)$ and let Q be the diagonal matrix $\operatorname{diag}(-\lambda_{d+1},\ldots,-\lambda_n)$. Then

$$-H_x(f) = \begin{bmatrix} P & 0 \\ 0 & Q \end{bmatrix}$$

in these coordinates. For $(x_1,\ldots,x_n) \in \mathbf{R}^n$ write $y = (x_1,\ldots,x_d)$ and $z = (x_{d+1},\ldots,x_n)$. Then the trajectories of $-\text{grad } f$ in these coordinates are the solution curves to the differential equation

10.11. $$\begin{aligned} y &= Py + F_1(y,z) \\ z &= Qz + F_2(y,z) \end{aligned}$$

where F_1 and F_2 are C^∞ and their Jacobian matrices ∂F_1 and ∂F_2 vanish at the origin (cf. [H] Chapter IX,§4). By reducing the neighbourhood W_x of x if necessary we may assume that F_1 and F_2 extend smoothly over $\mathbf{R}^n$ in such a way that there exist complete solution curves to 10.11 through every point $(y_0,z_0) \in \mathbf{R}^n$, given by

$$t \to (y_t,z_t)$$

say, for $t \in \mathbf{R}$ (see [H] IX 3 and 4). Then we have

10.12. $$\begin{aligned} y_t &= e^{Pt} y_0 + Y(t,y_0,z_0) \\ z_t &= e^{Qt} z_0 + Z(t,y_0,z_0) \end{aligned}$$

for all $t \in \mathbf{R}$, where Y,Z and their partial Jacobian matrices $\partial_{y_0,z_0} Y$ and $\partial_{y_0,z_0} Z$ vanish at the origin.

We want to show that if a point x does not lie in Σ_C then its path of steepest descent stays well away from C. If x is sufficiently close to Σ_C then it has a well-defined distance $d(x,\Sigma_C)$ from Σ_C. It is sufficient to show that near C this distance function is bounded away from zero along all paths of steepest descent not contained in Σ_C. We can do this by working in local coordinates near each $x \in C$.

The submanifold Σ_C is defined in the local coordinates (y,z) in W_x by $z = 0$. Therefore in the standard metric on $\mathbf{R}^n$ the distance from Σ_C is given by $\|z\|$. Moreover the coordinates were chosen so that the given Riemannian metric at x coincides with the standard inner product on $\mathbf{R}^n$. It follows that given any $\varepsilon > 0$ we may reduce W_x so that

10.13. $$(1+\varepsilon)^{-1}\|z\| \leq d((y,z),\Sigma_C) \leq (1+\varepsilon)\|z\|$$

everywhere in W_x.

We now need the following technical result.

10.14. <u>Lemma.</u> <u>There is a number</u> $b > 1$, <u>which depends only on the critical set</u> $C \in \mathbf{C}$, <u>such that the following property holds for every</u> $x \in C$. <u>If the neighbourhood</u> W_x <u>of</u> x <u>is taken sufficiently small and the extensions of</u> F_1 <u>and</u> F_2 <u>over</u> $\mathbf{R}^n$ <u>are chosen appropriately, then for every</u> $(y_0,z_0) \in \mathbf{R}^n$ <u>we have</u>

$$\|z_1\| \geq b\|z_0\|$$

where $z_1 = e^Q z_0 + Z(1,y_0,z_0)$ as at 10.12.

Proof. The gradient field of f is tangential to the submanifold Σ_C so $F_2(y,0) = 0$ whenever $(y,0)$ lies in W_x (see 10.11). Therefore the extension of F_2 to $\mathbf{R}^n$ can be chosen so that $F_2(y,0) = 0$ for all $y \in \mathbf{R}^d$. This implies that

$$Z(t,y_0,0) = 0 \quad \text{for all} \quad y_0 \in \mathbf{R}^d \quad \text{and} \quad t \in \mathbf{R}$$

(see 10.12).

Now for each $x \in C$ let c_x be the minimum eigenvalue of e^Q. Recall that

$$Q = \text{diag}\,(-\lambda_{d+1},\ldots,-\lambda_n)$$

where $\lambda_{d+1},\ldots,\lambda_n$ are the eigenvalues of the Hessian $H_x(f)$ restricted to $T_x\Sigma_C$, and that each of these eigenvalues is strictly negative. Hence $c_x > 1$. Let

$$c = \inf\{c_x \mid x \in C\};$$

since C is compact and c_x depends continuously on x it follows that $c > 1$. So we can choose $\theta > 0$ such that $c - \theta > 1$. Set $b = c - \theta$; then $b > 1$ and b depends only on C.

By 10.12 the partial Jacobian $\partial_{y_0,z_0} Z$ vanishes at the origin for all $t \in \mathbf{R}$. Hence by reducing the neighbourhood W_x and choosing the extensions of F_1 and F_2 appropriately we may assume that

$$\|\partial_{z_0} Z(1,y_0,z_0)\| \leq \theta \quad \text{for all} \quad (y_0,z_0) \in \mathbf{R}^n$$

(cf. [H] IX §4). It follows that

$$\|Z(1,y_0,z_0)\| \leq \theta\|z_0\| \quad \text{for all } (y_0,z_0) \in \mathbf{R}^n .$$

Since every eigenvalue of e^Q is at least c, for any $(y_0,z_0) \in \mathbf{R}$ we have

$$\begin{aligned}\|z_1\| &= \|e^{Qz_0} + Z(1,y_0,z_0)\| \\ &\geq c\|z_0\| - \theta\|z_0\| \\ &= b\|z_0\|.\end{aligned}$$

The result follows.

10.15. Corollary. The intersection of the Morse stratum S_C with a sufficiently small neighbourhood of C in X is contained in the minimising manifold Σ_C.

Proof. It follows from 10.13 and 10.14 that given $\varepsilon > 0$ there is a neighbourhood W_C of C such that if $\{x_t | t \geq 0\}$ is any path of steepest descent with $x_t \in W_C$ when $0 \leq t \leq 1$ then

$$d(x_1,\Sigma_C) \geq b(1+\varepsilon)^{-2}\, d(x_0,\Sigma_C)$$

where $b > 1$ is independent of ε. If ε is chosen sufficiently small we have

$$b(1+\varepsilon)^{-2} > 1.$$

By remark 10.9 there is a neighbourhood V_C of C in X such that if $x_0 \in V_C \cap S_C$ its entire path of steepest descent $\{x_t | t \geq 0\}$ is contained in W_C. Then for each $n \geq 1$

$$d(x_n,\Sigma_C) \geq (b(1+\varepsilon)^{-2})^n\, d(x_0,\Sigma_C).$$

But we may assume without loss of generality that $d(x,\Sigma_C)$ is bounded on

W_C. Hence we must have $d(x_0, \Sigma_C) = 0$, i.e. $x_0 \in \Sigma_C$. This shows that $V_C \cap S_C \subseteq \Sigma_C$.

From 10.8 and 10.15 we deduce that each stratum S_C coincides with Σ_C in a neighbourhood U_C of C, and hence that $S_C \cap U_C$ is smooth. But any point of S_C is mapped into $S_C \cap U_C$ by the diffeomorphism

$$x \to x_t$$

of S_C induced by flowing for some large time t along the gradient field of f. So we have the following

10.16. Lemma. *For each* $C \in \mathbf{C}$ *the stratum* S_C *is smooth. It coincides with the minimising manifold* Σ_C *in some neighbourhood of* C.

We have now proved that the subsets $\{S_C \mid C \in \mathbf{C}\}$ form a smooth stratification of X, and it remains only to prove one more result.

10.17. Lemma. *For each* $C \in \mathbf{C}$ *the inclusion* $C \to S_C$ *is an equivalence for Čech cohomology. More generally if a compact connected group* K *acts on* X *in such a way that the function* f *and the Riemannian metric on* X *are preserved by* K, *then each stratum* S_C *is K-invariant and the inclusions* $C \to S_C$ *are equivalences of equivariant cohomology.*

Proof. We need only consider the second statement. It is clear from the definition that the Morse strata $\{S_C \mid C \in \mathbf{C}\}$ are K-invariant.

For each sufficiently small $\delta \geq 0$,

$$N_\delta = \{x \in S_C \mid f(x) \leq f(C) + \delta\}$$

is a compact neighbourhood of C in S_C (cf. the proof of 10.8). The paths of steepest descent induce retractions of S_C onto each N_δ which respect the action of K. So each inclusion $N_\delta \times_K EK \to S_C \times_K EK$ is a cohomology equivalence. Also

$$\bigcap_{\delta>0} N_\delta = C.$$

So the continuity of Čech cohomology implies that the inclusion $C \to S_C$ is an equivalence of equivariant Čech cohomology (see [D] VIII 6.18). The only problem is that $X \times_K EK$ is not compact. This can be overcome by regarding EK as the union of compact manifolds which are cohomologically equivalent to EK up to arbitrarily large dimensions.

10.18. Remark. When f is nondegenerate in the sense of Bott each path of steepest descent under f converges to a unique critical point in X. Thus the strata retract onto the critical sets along the paths of steepest descent. This fails in general for minimally degenerate functions: there exist minimally degenerate functions with trajectories which 'spiral in' towards a critical subset without converging to a unique limit. This is why Čech cohomology is used above. However it is unlikely that the square of a moment map has such bad behaviour.

Part II. The algebraic approach.

§11. The basic idea.

In Part I a formula was obtained in good cases for the Betti numbers of the projective quotient variety associated in geometric invariant theory to a linear action of a complex reductive group G on a nonsingular complex projective variety X. The good cases occur when the stabiliser in G of every semistable point of X is finite. The quotient variety is then topologically the quotient X^{ss}/G of the set of semistable points by the group. The formula was obtained by employing the ideas of Morse theory and of symplectic geometry. We shall now approach the same problem using algebraic methods.

The basic idea common to both approaches is to associate to the group action a canonical stratification of the variety X. The unique open stratum of this stratification coincides with the set X^{ss} of semistable points of X (provided X^{ss} is nonempty) and the other strata are all G-invariant locally-closed nonsingular subvarieties of X. There then exist equivariant Morse-type inequalities relating the G-equivariant Betti numbers of X to those of the strata. It turns out that these inequalities are in fact equalities, i.e. that the stratification is equivariantly perfect over the rationals. From this an inductive formula can be derived for the equivariant Betti numbers of the semistable stratum X^{ss} which in good cases coincide with the ordinary Betti numbers of the quotient variety.

The difference between the two approaches lies in the way the stratification of X is defined. In Part I symplectic geometry was used to define a function f (the norm-square of the moment map) which induced a Morse stratification of X. In Part II the stratification will be defined purely algebraically. The main advantage of this method is that it applies to varieties defined over any algebraically closed field. On the other hand the approach of Part I generalises to Kähler and symplectic manifolds.

The algebraic definition of the stratification is based on work of Kempf. It also has close links with the paper [Ne] by Ness. Suppose that we are given a linear action of a reductive group G on any projective variety X, singular or nonsingular, defined over any algebraically closed field. Kempf shows that for each unstable point $x \in X$ there is a conjugacy class of virtual one-parameter groups of a certain parabolic subgroup of G which are 'most responsible' for the instability of x. (The terminology 'canonical destabilising flags' is also used). The stratum to which x belongs is determined by the conjugacy class of these virtual one-parameter subgroups in G. Over the complex field the stratification is the same as the one already defined in Part I.

Just as in Part I the indexing set $\mathbf{B}$ of the stratification may be described in terms of the weights of the representation of G which defines the action. An element $\beta \in \mathbf{B}$ may be thought of as the closest point to the origin of the convex hull of some nonempty set of weights, when the weights are regarded as elements of an appropriate normed space (see 12.8).

In §13 it is shown that if X is nonsingular then the strata S_β are also nonsingular and have the same structure as in the complex case. That is, for each β in the indexing set $\mathbf{B}$ there is a smooth locally-closed subvariety Y_β^{ss} of X, acted on by a parabolic subgroup P_β of G, such that

11.1. $$S_\beta \cong G \times_{P_\beta} Y_\beta^{ss} .$$

There is also a nonsingular closed subvariety Z_β of X and a locally trivial fibration

11.2. $$p_\beta : Y_\beta^{ss} \to Z_\beta^{ss}$$

whose fibres are all affine spaces. Here Z_β^{ss} is the set of semistable points of Z_β under the action of a reductive subgroup of P_β.

These results were precisely what was needed in Part I to show that the stratification $\{S_\beta | \beta \in \mathbf{B}\}$ is equivariantly perfect and hence to derive an inductive formula for the equivariant Betti numbers of X^{ss}. Thus the reader who is interested solely in complex algebraic varieties can avoid the detailed analytic arguments needed for symplectic and Kähler manifolds by using definitions and results from these two sections. It will be found that at times the algebraic method is neater while at others it is more elegant to argue analytically.

In §14 we shall see how the formulae for the Betti numbers of the quotient variety M can be refined to give the Hodge numbers as well. We use Deligne's extension of Hodge theory to complex varieties which are not necessarily compact and nonsingular.

In §15 an alternative method of obtaining the formulae is described, though without detailed proofs. This method was suggested by work of Harder and Narasimhan [H & N]. It uses the Weil conjectures which were established by Deligne. These enable one to calculate the Betti numbers of a nonsingular complex projective variety by counting the points of associated varieties defined over finite fields. In our case it is possible to count points by decomposing these varieties into strata and using 11.1 and 11.2. However the Weil conjectures only apply when the quotient variety is nonsingular.

Finally in §16 some examples of stratifications and of calculating the Betti numbers of quotients are considered in detail. The first example is given by the action of SL(2) on the space $\mathbf{P}_n$ of binary forms of degree n, which can be identified with the space of unordered sets of points on the complex projective line $\mathbf{P}_1$. We also consider the space $(\mathbf{P}_1)^n$ of ordered sets of points on $\mathbf{P}_1$. These have been used as examples throughout Part I. The good cases occur when n is odd, and then the Hodge numbers of the quotient varieties M are given by

$$h^{p,p} = [1 + 1/2 \min (p,n-3-p)]$$

for the case of unordered points, and

$$h^{p,p} = 1 + (n-1) + \binom{n-1}{2} + \dots + \binom{n-1}{\min(p,n-3-p)}$$

for ordered points. The Hodge numbers $h^{p,q}$ with $p \neq q$ all vanish. Then we generalise $(\mathbf{P}_1)^n$ to an arbitrary product of Grassmannians. That is, we consider for any m the diagonal action of $SL(m)$ on a product of Grassmannians $G(\ell_i,m)$ where $G(\ell,m)$ denotes the Grassmannian of ℓ-dimensional subspaces of $\mathbf{C}^m$. The good cases occur when m is coprime to $\sum_i \ell_i$. The associated stratification is described in Proposition 16.9 and it is shown how in good cases this provides an inductive formula for the equivariant Betti numbers of the semistable stratum in terms of the equivariant Betti numbers of the semistable strata of products of the same form but with smaller values of m. Explicit calculations are made for some products of $\mathbf{P}_2$.

One reason for studying products of Grassmannians in depth is that it is possible to rederive the formulae obtained in [H & N] and [A & B] for the Betti numbers of moduli spaces of vector bundles over Riemann surfaces by applying the results of these notes to subvarieties of products of Grassmannians (see [Ki3]).

§12. Stratifications over arbitrary algebraically closed fields

Let k be an algebraically closed field. Suppose that X is a projective k-variety acted on linearly by a reductive k-group G. In this section we shall define a stratification of X which generalises the definition given in Part I for the case when X is nonsingular and k is the field of complex numbers.

The set X^{ss} of semistable points of X under the action will form one stratum of the stratification. To define the others we shall use work of Kempf as expounded in a paper by Hesselink (see [K] and [Hes]). Kempf associates to each unstable point x of X a conjugacy class of virtual one-parameter subgroups in a parabolic subgroup of G. These are the ones 'most responsible' for the instability of the point x. The stratum to which x belongs will be determined by the conjugacy class of these virtual one-parameter subgroups in G. We shall find that each stratum S_β can be described in the form

$$S_\beta = GY_\beta^{ss}$$

where Y_β^{ss} is a locally-closed subvariety of X, itself defined in terms of the semistable points of a smaller variety under the action of a subgroup of G. From this it will be obvious that the stratification coincides with the one defined in Part I in the complex nonsingular case.

First we shall review briefly Hesselink's definitions and results and relate them to what we have already done in the complex case: this is completed in lemma 12.13. Note that in [Hes] arbitrary ground fields are considered. We shall restrict ourselves to algebraically closed fields for the sake of simplicity.

Remark. The definition of the stratification given at 12.14 makes sense when k is any field. There is also a stratification of the variety $X \times_k K$ defined over the algebraic closure K of k. When k is perfect it follows from [Hes] that this last stratification is defined over k and coincides with the first stratification on X. However this fails in general (see [Hes] example 5.6). In §15 where finite fields occur it will be necessary to avoid certain characteristics where things go wrong.

In [Hes] Hesselink studies reductive group actions on affine pointed varieties. We shall apply his results to the action of G on the affine cone $X^* \subseteq k^{n+1}$ on X. For each nonzero x^* in X^* and one-parameter subgroup $\lambda: k^* \to G$ of G Hesselink defines a "measure of instability" $m(x^*;\lambda)$. This depends only on the point x in X determined by x^* and hence can also be written as $m(x;\lambda)$. The following two facts determine $m(x;\lambda)$ for every x and λ.

12.1. If $\lambda: k^* \to GL(n+1)$ is given by

$$z \to \operatorname{diag}(z^{r_0},\dots,z^{r_n})$$

with $r_0,\dots,r_n \in \mathbb{Z}$ then

$$m(x;\lambda) = \min\{r_j \mid x_j \neq 0\}$$

<u>if this is non-negative and</u>

$$m(x;\lambda) = 0$$

<u>otherwise.</u> <u>Also</u> $m(x;g\lambda g^{-1}) = m(gx;\lambda)$ <u>for any</u> $g \in G$.

12.2. <u>Definition.</u> $x \in X$ is <u>unstable</u> for the action of G if $m(x;\lambda) > 0$ for some one-parameter subgroup λ of G.

In [M] theorem 2.1, Mumford proves that

12.3. $x \in X$ <u>is semistable if and only if</u> $m(x;\lambda) \leq 0$ <u>for every one-parameter subgroup</u> λ <u>of</u> G; <u>that is, if and only if</u> x <u>is not unstable.</u>

12.4. <u>Definitions</u> ([Hes] §1). Let $Y(G)$ be the set of one-parameter subgroups $\lambda: k^* \to G$ of G, and let $M(G)$ be the quotient of the product of $Y(G)$ with the natural numbers by the equivalence relation $\sim$ such that $(\lambda,\ell) \sim (\mu,m)$ if λ and μ satisfy the formula

$$\lambda(t^m) = \mu(t^\ell).$$

If T is a torus then $Y(T)$ is a free $\mathbf{Z}$-module of finite rank and $M(T)$ is a Q-vector space. Moreover there is a natural correspondence between one-parameter subgroups of a torus T over the complex field and lattice points in the Lie algebra $\mathbf{t}$ of its maximal compact subgroup. Hence in this case $M(T)$ may be identified with the rational points of $\mathbf{t}$.

The adjoint action of G on $Y(G)$ extends to an action on $M(G)$. Let q be a norm on $M(G)$ as defined in [Hes] §1; that is, q is a G-invariant map from $M(G)$ to $\mathbf{Q}$ which restricts to a quadratic form on $M(T)$ for any torus $T \subseteq G$. If T is a maximal torus of G a norm on $M(G)$ is the square of an inner product on $M(T)$ invariant under the Weyl group, and any such inner product determines a unique norm on $M(G)$. When $k = \mathbf{C}$ any invariant rational inner product on the Lie algebra of a maximal compact subgroup of G induces a norm on $M(G)$ in Hesselink's sense.

12.5. Definitions ([Hes] 4.1). For any $x \in X$ let

$$q_G^{-1}(x) = \inf\{q(\lambda) \mid \lambda \in M(G),\ m(x,\lambda) \geq 1\}$$

and

$$\Lambda_G(x) = \{\lambda \in M(G) \mid m(x;\lambda) \geq 1,\ q(\lambda) = q_G^{-1}(x)\}.$$

Thus x is unstable if and only if $q_G^{-1}(x) < \infty$ or equivalently $\Lambda_G(X) \neq \emptyset$. The definition of $m(x;\lambda)$ can be extended uniquely over all λ in $M(G)$ to satisfy $m(x;r\lambda) = rm(x;\lambda)$ for every $r \in \mathbf{Q}$.

The set $\Lambda_G(x)$ will be used to determine the stratum to which the point $x \in X$ belongs.

Let T be a maximal torus of G. As when $k = \mathbf{C}$ the representation of T on k^{n+1} splits as the sum of scalar representations given by characters $\alpha_0,\ldots,\alpha_n$ say. These α_i are elements of the dual of $M(T)$ and may be identified with elements of $M(T)$ by using the inner product on $M(T)$ whose square is the norm q.

Fix $x = (x_0 : \ldots : x_n) \in X$ and let β be the closest point to the origin for the norm q of the convex hull $C(x)$ of the set $\{\alpha_i \mid x_i \neq 0\}$ in the $\mathbb{Q}$-vector space $M(T)$. Then

$$(\xi - \beta).\beta \geq 0, \text{ i.e. } \xi.\beta \geq q(\beta)$$

for all $\xi \in C(x)$, where . denotes the inner product on $M(T)$ whose square is q. In Part I this point β indexed the stratum containing x. The next two lemmas show how β is related to the set $\Lambda_T(x)$.

12.6. <u>Lemma</u>. <u>If</u> $\beta \neq 0$ <u>then</u> $\Lambda_T(x) = \{\beta/q(\beta)\}$.

<u>Proof</u>. By 12.1 if $\lambda \in M(T)$ then $m(x;\lambda) = \min\{\alpha_i.\lambda \mid x_i \neq 0\}$ if this is nonnegative and $m(x;\lambda) = 0$ otherwise. Therefore $m(x;\lambda) \geq 1$ if and only if $\lambda.\alpha_i \geq 1$ for every i such that $x_i \neq 0$. But if $x_i \neq 0$ then $\alpha_i.\beta \geq q(\beta)$ by the choice of β. Therefore $\beta/q(\beta) . \alpha_i \geq 1$ for such i. Moreover if λ satisfies $\lambda.\alpha_i \geq 1$ whenever $x_i \neq 0$ then $\lambda.\beta \geq 1$ since β lies in the convex hull of the set $\{\alpha_i \mid x_i \neq 0\}$. This means that $q(\lambda)q(\beta) \geq (\lambda.\beta)^2 \geq 1$ with equality if and only if $\lambda = \beta/q(\beta)$. Thus it follows straight from definition 12.5 that $q_G^{-1}(x) = q(\beta)^{-1}$ and that the set $\Lambda_T(x)$ consists of the single point $\beta/q(\beta)$.

12.7. <u>Lemma</u>. <u>If</u> $\beta = 0$ <u>then</u> $\Lambda_T(x) = \emptyset$.

<u>Proof</u>. If $\Lambda_T(x) \neq \emptyset$ then there is some $\lambda \in M(T)$ such that $m(x;\lambda) \geq 1$ and hence such that $\lambda.\alpha_i \geq 1$ whenever $x_i \neq 0$. This implies that $0 \notin \mathrm{Conv}\{\alpha_i \mid x_i \neq 0\}$ and hence that $\beta \neq 0$. The result follows.

Thus the set $\Lambda_T(x)$ determines and is determined by the point β.

12.8. Definition. Call the closest point to 0 of the convex hull in $M(T)$ of any nonempty subset of $\{\alpha_0,\ldots,\alpha_n\}$ a minimal combination of weights. Let **B** be the set of all minimal combinations of weights lying in some positive Weyl chamber (i.e. some convex fundamental domain for the action of the Weyl group on $M(T)$). **B** will be the indexing set for the stratification.

12.9. Definition ([Hes] §4). A subgroup H of G is optimal for x if $q_H^{-1}(x) = q_G^{-1}(x)$.

It is clear from definition 12.5 that if H is optimal then $\Lambda_H(x) = M(H) \cap \Lambda_G(x)$ and that $\Lambda_H(x)$ is nonempty precisely when $\Lambda_G(x)$ is nonempty. By [Hes] 4.4(b) there is always a maximal torus T' of G which is optimal for x, and by [Hes] 4.3 $T' = g^{-1}Tg$ for some $g \in G$ where T is the fixed maximal torus of G. This implies that

12.10. For every $x \in X$ there exists some $g \in G$ such that T is optimal for gx.

Next note that G acts on itself by conjugation and hence G becomes an affine pointed G-variety. So we can make the following definition.

12.11. Definition. If $\lambda \in M(G)$ let

$$P_\lambda = \{g \in G \mid m(g;\lambda) \geq 0\}.$$

Clearly if $r > 0$ is rational then $P_\lambda = P_{r\lambda}$. Moreover if $\lambda: k^* \to G$ is actually a one-parameter subgroup of G then P_λ consists of those $g \in G$ such that $\lim_{t \to 0} \lambda(t)g\lambda(t)^{-1}$ exists in G. Then lemma 5.1(a) of [Hes] or an argument similar to that of lemma 6.9 in Part I shows that

12.12. P_λ <u>is a parabolic subgroup of</u> G <u>for each</u> $\lambda \in M(G)$.

The main result needed from Kempf's work can now be stated.

12.13. <u>Lemma.</u> (i) <u>For each unstable</u> $x \in X$ <u>there is a unique parabolic subgroup</u> $P(x)$ <u>of</u> G <u>such that</u> $P(x) = P_\lambda$ <u>for all</u> λ <u>in</u> $\Lambda_G(x)$.

(ii) $\Lambda_G(x)$ <u>is a single</u> $P(x)$ <u>orbit under the adjoint action of</u> G <u>on</u> $M(G)$.

(iii) <u>If</u> $\lambda \in \Lambda_G(x)$ <u>then</u> $g^{-1}\lambda g$ <u>also lies in</u> $\Lambda_G(x)$ <u>if and only if</u> $g \in P(x)$. <u>In particular</u> $P(x)$ <u>contains the stabiliser of</u> x <u>in</u> G.

(iv) $\Lambda_G(x) \subseteq M(P(x))$.

(v) <u>If</u> T <u>is optimal for</u> x <u>and</u> $\Lambda_T(x) = \{\beta/q(\beta)\}$ <u>then</u> $P(x) = P_\beta$.

<u>Proof.</u> (i) and (ii) are [Hes] theorem 5.2 applied to any nonzero $x^* \in X^*$ lying over x. If $\lambda: k^* \to G$ is a one-parameter subgroup of G and if $g \in G$ is such that $g^{-1}\lambda g = \lambda$, then g commutes with every element of $\lambda(k^*)$ so $m(g;\lambda) = 0$ and hence $g \in P_\lambda$. The first part of (iii) follows from this together with (i) and (ii). The second part is an immediate consequence

since $m(x;g^{-1}\lambda g) = m(gx;\lambda)$ by 12.1. (iv) also follows because if $\lambda \in \Lambda_G(x)$ then $r\lambda \in Y(G)$ for some positive integer r. Since $r\lambda$ commutes with λ it represents a one-parameter subgroup of $P_\lambda = P(x)$ and so $\lambda \in M(P(x))$. Finally if T is optimal for x and $\Lambda_T(x) = \beta/q(\beta)$ then $\beta/q(\beta) \in \Lambda_G(x)$, so $P(x) = P(\beta/q(\beta)) = P_\beta$ by (i) and 12.11.

This lemma completes the review of the results needed from [Hes].

12.14. <u>Definition.</u> For each nonzero $\beta \in M(T)$ let

$$S_\beta = G\{x \in X \mid \beta/q(\beta) \in \Lambda_G(x)\}$$

and let

$$S_0 = G\{x \in X \mid \Lambda_G(x) = \emptyset\}.$$

Then by 12.3 $S_0 = X^{ss}$. Also $\beta/q(\beta) \in \Lambda_G(x)$ if and only if T is optimal for x and $\Lambda_T(x) = \{\beta/q(\beta)\}$ by 12.6 and 12.7.

12.15. <u>Lemma.</u> X <u>is the disjoint union of the subsets</u>

$$\{S_\beta \mid \beta \in B\}.$$

<u>Proof.</u> Suppose that $x \in X$ is unstable, i.e. that $\Lambda_G(x) \neq \emptyset$. By 12.10 there is some $g \in G$ such that T is optimal for gx. By 12.6 and 12.7 $\Lambda_T(gx) = \{\beta/q(\beta)\}$ where $\beta \neq 0$ is the closest point to 0 of $\mathrm{Conv}\{\alpha_i \mid (gx)_i \neq 0\}$. Therefore

$$\bigcup_\beta S_\beta = X,$$

where β runs over all minimal combinations of weights.

Since x is unstable if $\Lambda_T(gx) \neq \emptyset$ for any $g \in G$,

$$S_0 \cap \bigcup_{\beta \neq 0} S_\beta = \emptyset.$$

If β and β' are nonzero and the intersection $S_\beta \cap S_{\beta'}$ is nonempty then there exist $x \in X$ and $g \in G$ such that both $\beta/q(\beta)$ and $\mathrm{Ad}g(\beta'/q(\beta'))$ lie in $\Lambda_G(x)$. Therefore by 12.13 (ii) $\beta/q(\beta)$ and $\beta'/q(\beta')$ are equivalent under the adjoint representation of G on $M(G)$. This implies that $q(\beta) = q(\beta')$ so β and β' are also equivalent. As β and β' lie in $M(T)$ it follows that β and β' lie in the same orbit of the Weyl group in $M(T)$.

Conversely suppose that β and β' are equivalent under the action of the Weyl group, so that there is some $g \in G$ normalising T such that $\beta' = \mathrm{Ad}g(\beta)$. Then for any x we have $\beta/q(\beta) \in \Lambda_G(x)$ if and only if $\beta'/q(\beta') \in \Lambda_G(gx)$, so $S_\beta = S_{\beta'}$. The result follows.

Write $\beta' > \beta$ if $q(\beta') > q(\beta)$. In order to show that we have a stratification of X (in the Zariski topology) it now suffices to show that

12.16. <u>Lemma.</u> $\overline{S}_\beta \subseteq \bigcup_{\beta' \geq \beta} S_{\beta'}$ <u>for each</u> $\beta \in \mathbf{B}$.

<u>Proof.</u> For each $\beta \in \mathbf{B}$ let

$$W_\beta = \{x \in X \mid x_i = 0 \text{ if } \alpha_i . \beta < q(\beta)\}.$$

By 12.6, 12.7 and the preceding remarks the stratum S_β consists of all points of the form gx such that T is optimal for x and β is the closest point to 0 of $\mathrm{Conv}\{\alpha_i \mid x_i \neq 0\}$. This implies that

$$S_\beta \subseteq GW_\beta$$

for each $\beta \in \mathbf{B}$. It is easy to check that W_β is invariant under P_β (see

12.23 below noting that $W_\beta = \overline{Y}_\beta$ when X is projective space). By a standard argument using the completeness of G/P_β (see e.g. [B] 11.9 (i), [Hes] 6.3 or theorem 13.7 below) it follows that GW_β is closed in X so that

$$\overline{S}_\beta \subseteq GW_\beta .$$

Suppose $x \in W_\beta$ and let β' be the closest point to 0 of $\mathrm{Conv}\{\alpha_i \mid x_i \neq 0\}$. Then either $\beta' = \beta$ or $q(\beta') > q(\beta)$. If T is optimal for x then $x \in S_{\beta'}$ and otherwise by the definition of optimality we must have $x \in S_{\beta''}$ for some β'' with $q(\beta'') > q(\beta') \geq q(\beta)$. Therefore

12.17. if $x \in W_\beta$ then either T is optimal for x and β is the closest point to 0 of $\mathrm{Conv}\{\alpha_i \mid x_i \neq 0\}$ or there is some $\beta' > \beta$ such that x belongs to $S_{\beta'}$.

Hence

$$\overline{S}_\beta \subseteq GW_\beta \subseteq \bigcup_{\beta' \geq \beta} S_{\beta'}$$

so the proof is complete.

This lemma shows that the subsets $\{S_\beta \mid \beta \in \mathbf{B}\}$ form a stratification of X in the sense of definition 2.11, and in particular

$$S_\beta = \overline{S}_\beta - \bigcup_{\beta' > \beta} \overline{S}_{\beta'}$$

is open in its closure $\overline{S}_\beta$ for each $\beta \in \mathbf{B}$.

We next want to describe the stratum S_β in such a way that it is clear that when $k = \mathbf{C}$ this stratification coincides with the one defined in Part I.

12.18. Definition. Let

$$Z_\beta = \{(x_0:\ldots:x_n) \in X \mid x_j = 0 \text{ if } \alpha_j.\beta \neq q(\beta)\}$$

and let

$$Y_\beta = \{(x_0:\ldots:x_n) \in X \mid x_j = 0 \text{ if } \alpha_j.\beta < q(\beta),\ x_j \neq 0 \text{ for some } j \text{ with } \alpha_j.\beta = q(\beta)\}.$$

Z_β is a closed subvariety of X and Y_β is a locally-closed subvariety. Define $p_\beta : Y_\beta \to Z_\beta$ by

$$p_\beta(x_0:\ldots:x_n) = (x_0':\ldots:x_n')$$

where $x_j' = x_j$ if $\alpha_j.\beta = q(\beta)$ and $x_j' = 0$ otherwise. This is well defined as a map $Y_\beta \to Z_\beta$ since if $y \in Y_\beta$ then $p_\beta(y) \in \overline{Gy}$ and in particular lies in X. Let $\mathrm{Stab}\beta$ be the stabiliser of β under the adjoint action of G on $M(G)$. Then $\mathrm{Stab}\beta$ is a reductive subgroup of G which acts on Z_β.

The definitions of Z_β, Y_β and p_β depend only on β. They are independent of the choice of coordinates, and indeed of the maximal torus T chosen except that β must lie in $M(T)$. Moreover by 6.5 when $k = \mathbf{C}$ and X is nonsingular they coincide with the definitions made in Part I.

12.19. Lemma. If $x \in Z_\beta$ then $\mathrm{Stab}\beta$ is optimal for x.

Proof. If $x \in Z_\beta$ then β fixes x so $\beta \in M(P(x))$ by 12.13 (iii). Also $\Lambda_G(x) \subseteq M(P(x))$ by 12.13 (iv) so that if $\lambda \in \Lambda_G(x)$ there is some $p \in P(x)$ such that $p\lambda p^{-1}$ and β commute. But this implies that $p\lambda p^{-1} \in M(\mathrm{Stab}\beta) \cap \Lambda_G(x)$ by 12.13 (ii) so $\mathrm{Stab}\beta$ is optimal for x as required.

Note that if $x \in Z_\beta$ then by definition

$$m(x;\beta) = \min\{\alpha_i.\beta \mid x_i \neq 0\} = q(\beta).$$

Thus in particular when $\beta \neq 0$ no point in Z_β is semistable. However there is an open subset of Z_β whose elements are unstable 'only insofar as β makes them unstable'. The neatest definition of this subset is the following.

12.20. Definition. Let Z_β^{ss} be the subset of Z_β consisting of those $x \in Z_\beta$ such that $\beta/q(\beta) \in \Lambda_G(x)$.

Since $\mathrm{Stab}\beta$ is optimal for x the condition that $\beta/q(\beta) \in \Lambda_G(x)$ is equivalent to the condition that

$$m(x;\lambda) \leq \lambda.\beta \quad \text{for every } \lambda \in M(\mathrm{Stab}\beta).$$

Note that $\lambda.\beta$ makes sense for $\lambda \in M(\mathrm{Stab}\beta)$ because there is some maximal torus T' such that λ and β both lie in $M(T')$ and there is a unique inner product on $M(T')$ whose square is the norm q.

Let Y_β^{ss} be the inverse image of Z_β^{ss} under the map $p_\beta : Y_\beta \to Z_\beta$ defined at 12.18.

12.21. Remark. It is not hard to give alternative definitions of Z_β^{ss} and Y_β^{ss} directly in terms of semistability (cf. 8.11). One can show that there is a unique connected reductive subgroup G_β of $\mathrm{Stab}\beta$ such that $M(G_\beta) = \{\lambda \in M(\mathrm{Stab}\beta) \mid \lambda.\beta = 0\}$. Then Z_β^{ss} consists precisely of those $x \in Z_\beta$ which are semistable under the action of G_β on Z_β via the restriction of the homomorphism $G \to GL(n+1)$ to G_β. This is easily seen by using lemmas 12.6 and 12.7 together with 12.3.

Alternatively there exists a positive integer r such that when $M(T)$ is identified with its dual $r\beta$ corresponds to a character of T which extends to a character χ of $\mathrm{Stab}\beta$. Then the action of $\mathrm{Stab}\beta$ on Z_β is linearised with respect to the rth tensor power of the hyperplane bundle by the rth tensor power of the homomorphism $G \to GL(n+1)$ multiplied by the character χ^{-1}. It is not hard to check that a point x lies in Z_β^{ss} if and only if x is semistable for this linear action of $\mathrm{Stab}\beta$ on Z_β.

It is clear from the definition that

12.22. Z_β^{ss} <u>is invariant under</u> $\mathrm{Stab}\beta$,

and it follows that

12.23. Y_β <u>and</u> Y_β^{ss} <u>are invariant under</u> P_β.

The proof is essentially that of 6.10. It depends on two facts: firstly that if $\lambda: k^* \to T$ is any one-parameter subgroup which is a positive scalar multiple of β in $M(T)$ then $\lim_{t\to 0} \lambda(t)g\lambda(t)^{-1}$ exists in G, and secondly that if $y \in Y_\beta$ then

$$p_\beta(y) = \lim_{t\to 0} \lambda(t)y$$

for any such $\lambda: k^* \to T$.

Our aim is now to show that $S_\beta = GY_\beta^{ss}$. For this the following lemma is needed.

12.24. Lemma. Suppose that $\beta \neq 0$. If $y \in Y_\beta$ and $x = p_\beta(y)$ then the following are equivalent.

(i) T is optimal for y and $\Lambda_T(y) = \{\beta/q(\beta)\}$.

(ii) $y \in S_\beta$.

(iii) $y \in Y_\beta^{ss}$.

(iv) $x \in Z_\beta^{ss}$.

(v) $x \in S_\beta$.

Proof. (iii) and (iv) are equivalent by definition, while (i) implies (ii) by definition 12.14 and the converse follows from 12.17 since $Y_\beta \subseteq W_\beta$. By 12.17 again if $y \notin S_\beta$ then $y \in S_{\beta'}$ for some β' satisfying $q(\beta') > q(\beta)$. Then $x \in \overline{S}_{\beta'}$ since $x \in \overline{Gy}$, and by lemma 12.16 this implies that $x \notin S_\beta$. Therefore (v) implies (ii). It follows straight from the definitions that (iv) implies (v).

Finally suppose that $x \notin Z_\beta^{ss}$. Since T is a maximal torus of $\mathrm{Stab}\beta$ there is some $s \in \mathrm{Stab}\beta$ such that T is optimal for sx. By 12.6 β is not the closest point to 0 of $\mathrm{Conv}\{\alpha_i \mid (sx)_i \neq 0\}$. Moreover $(sx)_i \neq 0$ if and only if both $(sy)_i \neq 0$ and $\alpha_i.\beta = q(\beta)$ because $p_\beta(sy) = sx$ (see definition 12.18). So it follows from the geometry of convex sets that β is not the closest point to 0 of $\mathrm{Conv}\{\alpha_i \mid (sy)_i \neq 0\}$. (This is best seen by drawing a picture). Thus by 12.6 and 12.7 $\Lambda_T(sy) \neq \{\beta/q(\beta)\}$ and hence by

12.17 $sy \in S_{\beta'}$ for some $\beta' > \beta$. So $y \notin S_\beta$. Thus (ii) implies (iv) and the proof is complete.

12.25. Corollary. If $\beta \neq 0$ then $y \in Y_\beta^{ss}$ if and only if T is optimal for y and $\Lambda_T(y) = \{\beta/q(\beta)\}$, or equivalently if and only if $\beta/q(\beta) \in \Lambda_G(x)$. Thus $S_\beta = GY_\beta^{ss}$ for any $\beta \in \mathbf{B}$.

Proof. It is obvious that $GY_0^{ss} = X^{ss} = S_0$. If $\beta \neq 0$ and $\Lambda_T(y) = \{\beta/q(\beta)\}$ then by 12.6 and 12.7 β is the closest point to 0 of $\mathrm{Conv}\{\alpha_i \mid y_i \neq 0\}$. Thus $y \in Y_\beta$ so the result follows straight from lemma 12.24.

We have now proved the following theorem.

12.26. Theorem. Let $X \subseteq \mathbf{P}_n$ be a projective variety over k and let G be a reductive k-group. Fix a norm q on the space $M(G)$ of virtual one-parameter subgroups of G. Then to any linear action of G on X there is associated a stratification $\{S_\beta \mid \beta \in \mathbf{B}\}$ of X by G-invariant locally-closed subvarieties described as follows. If T is a maximal torus of G the indices $\beta \in \mathbf{B}$ are minimal combinations of weights in a fixed Weyl chamber of $M(T)$ and

$$S_0 = X^{ss}$$

while if $\beta \neq 0$

$$S_\beta = GY_\beta^{ss}$$

where

$$Y_{\beta}^{ss} = \{x \in X \mid \beta/q(\beta) \in \Lambda_G(x)\}.$$

When $k = \mathbb{C}$ *and* X *is nonsingular the strata* S_{β} *and the subvarieties* Y_{β}^{ss} *coincide with those defined in Part I.*

§13. The strata of a nonsingular variety

Now suppose that X is a nonsingular projective variety over k. In this section we shall see that the strata $\{S_\beta \mid \beta \in B\}$ of the stratification associated in §12 to the action of a reductive group G are all nonsingular subvarieties of X. To prove this we shall show firstly that the subvarieties Z_β and Y_β defined at 12.20 are all nonsingular and secondly that each stratum S_β is isomorphic to $G \times_{P_\beta} Y_\beta^{ss}$. In addition we shall see that each morphism $p_\beta: Y_\beta^{ss} \to Z_\beta^{ss}$ is an algebraic locally trivial fibration such that every fibre is an affine space.

The following facts about linear actions of the multiplicative group k^* on nonsingular projective varieties such as X will be needed. These are due to Bialynicki-Birula (see [B-B] especially theorem 4.3). We shall apply them to certain one-parameter subgroups of G.

13.1. Suppose k^* acts linearly on X. Then the set of fixed points is a finite disjoint union of closed connected nonsingular subvarieties of X; let Z be one of these. For every $x \in X$ the morphism $k^* \to X$ given by $t \to tx$ extends uniquely to a morphism $k \to X$; the image of 0 will be denoted by $\lim_{t\to 0} tx$. Let Y consist of all $x \in X$ such that $\lim_{t\to 0} tx$ lies in Z. Then Y is a connected locally-closed nonsingular subvariety of X and the map $p: Y \to Z$ defined by

$$p(x) = \lim_{t \to 0} tx$$

is an algebraic locally trivial fibration with fibre some affine space over k.

13.2. Corollary. For each $\beta \in \mathbf{B}$ the subvarieties Y_β and Z_β defined at 12.18 are nonsingular. The morphism

$$p_\beta : Y_\beta \to Z_\beta$$

is an algebraic locally trivial fibration whose fibre at any point is an affine space. The same is therefore true of its restriction

$$p_\beta : Y_\beta^{ss} \to Z_\beta^{ss}$$

to the open subset Y_β^{ss} of Y_β.

Proof. Fix $\beta \in \mathbf{B}$ and let $r > 0$ be an integer such that $r\beta \in M(T)$ corresponds to a one-parameter subgroup of T. This one-parameter subgroup acts on X as

$$t \to \operatorname{diag}(t^{r\alpha_0 \cdot \beta}, \dots, t^{r\alpha_n \cdot \beta})$$

where $\alpha_0, \dots, \alpha_n$ are the weights of the representation of T on k^{n+1}. The definition of Z_β and Y_β shows that Z_β is a union of components of the fixed point set of this action and that $x \in Y_\beta$ if and only if $\lim_{t \to 0} tx \in Z_\beta$, in which case this limit coincides with $p_\beta(x)$. So the result is an immediate consequence of 13.1.

Now we want to show that each stratum S_β is isomorphic to $G \times_{P_\beta} Y_\beta^{ss}$ where P_β is the parabolic subgroup of G defined in 12.11. For simplicity we shall assume that the homomorphism $\phi: G \to GL(n+1)$ which defines the action of G on X is faithful. The general result follows immediately from this except that P_β must be replaced by $\phi^{-1}(\phi(P_\beta))$, which is also a parabolic subgroup of G.

13.3. <u>Definition</u> ([B] 3.3). Let $\mathbf{g}$ be the Lie algebra of the k-group G and for each $\beta \in \mathbf{B}$ let $\mathbf{p}_\beta$ be the Lie algebra of the parabolic subgroup P_β.

As a k-vector space $\mathbf{g}$ is just the tangent space to the group G at the origin. The action of G on X induces a k-linear map

$$\xi \to \xi_x$$

from $\mathbf{g}$ to the Zariski tangent space T_xX for each $x \in X$.

13.4. <u>Lemma. Suppose</u> G <u>is a subgroup of</u> $GL(n+1)$. <u>If</u> $x \in Y_\beta^{ss}$ <u>then</u>

$$\{g \in G \mid gx \in Y_\beta^{ss}\} = P_\beta$$

<u>and</u>

$$\{\xi \in \mathbf{g} \mid \xi_x \in T_x(Y_\beta^{ss})\} = \mathbf{p}_\beta \ .$$

<u>Proof</u> (compare the proof of lemma 6.15). By 12.23 Y_β^{ss} is invariant under P_β so $P_\beta \subseteq \{g \in G \mid gx \in Y_\beta^{ss}\}$ and $\mathbf{p}_\beta \subseteq \{\xi \in \mathbf{g} \mid \xi_x \in T_xY_\beta^{ss}\}$.

By 12.24 $x \in Y_\beta^{ss}$ if and only if T is optimal for x and $\Lambda_T(x) = \{\beta/q(\beta)\}$. Suppose that x and gx both lie in Y_β^{ss} for some

$g \in G$. Then $\beta/q(\beta) \in \Lambda_G(gx)$ so that $\beta/q(\beta)$ and $Ad(g^{-1})\beta/q(\beta)$ both lie in $\Lambda_G(x)$. Therefore $g \in P_\beta$ by 12.13 (iii).

It remains to show that $\{\xi \in \mathbf{g} \mid \xi_x \in T_x Y_\beta^{ss}\} \subseteq \mathbf{p}_\beta$. As in the proof of 13.2 if r is any positive integer such that $r\beta$ is a one-parameter subgroup of T then $r\beta$ acts on X as

$$t \to \operatorname{diag}(t^{r\alpha_0 \cdot \beta}, \dots, t^{r\alpha_n \cdot \beta}).$$

By 12.11 the subgroup P_β consists of all $g \in G$ such that $(r\beta(t))g(r\beta(t))^{-1}$ tends to some limit in G as $t \in k^*$ tends to 0. Hence an element g of G lies in P_β if and only if it is of the form $g = (g_{ij})$ with $g_{ij} = 0$ when $\alpha_i \cdot \beta < \alpha_j \cdot \beta$.

Let $\mathbf{g} = \mathbf{t} + \sum_{\alpha} \mathbf{g}^\alpha$ be the root space decomposition of $\mathbf{g}$ with respect to the Lie algebra $\mathbf{t}$ of the maximal torus T (see [B] theorem 13.18). If $\xi \in \mathbf{g}^\alpha$ has a nonzero ij-component then as $[\eta, \xi] = \alpha(\eta)\xi$ for all $\eta \in \mathbf{t}$ it follows that $\alpha = \alpha_i - \alpha_j$. So $\mathbf{g}^\alpha \subseteq \mathbf{p}_\beta$ whenever $\alpha \cdot \beta \geq 0$.

Hence it suffices to show that if $\xi \in \sum_{\alpha \cdot \beta < 0} \mathbf{g}^\alpha$ and $\xi_x \in T_x Y_\beta^{ss}$ then $\xi \in \mathbf{p}_\beta$.

Let V_+ (respectively V_0, V_-) be the sum of all the subspaces of k^{n+1} on which T acts as multiplication by some character α_i with $\alpha_i \cdot \beta > q(\beta)$ (respectively $\alpha_i \cdot \beta = q(\beta)$, $\alpha_i \cdot \beta < q(\beta)$). Then any element of $\sum_{\alpha \cdot \beta < 0} \mathbf{g}^\alpha$

is of block form $\begin{bmatrix} a & 0 & 0 \\ b & 0 & 0 \\ c & d & e \end{bmatrix}$ with respect to the decomposition of k^{n+1} as $V_+ \oplus V_0 \oplus V_-$.

If $x \in Z_\beta^{ss}$ then x is represented by a vector of the form $(0,v,0)$ in $k^{n+1} = V_+ \oplus V_0 \oplus V_-$. We have

$$\begin{bmatrix} a & 0 & 0 \\ b & 0 & 0 \\ c & d & e \end{bmatrix} \begin{bmatrix} 0 \\ v \\ 0 \end{bmatrix} = \begin{bmatrix} 0 \\ 0 \\ dv \end{bmatrix}$$

and so by the definition of Y_β^{ss} if $\xi_x \in T_x Y_\beta^{ss}$ then $dv = 0$ and hence $\xi_x = 0$. But this means that ξ is contained in the Lie algebra of the stabiliser of x in G, and by the first part of the lemma the stabiliser of x is contained in P_β. Therefore $\xi \in \mathbf{p}_\beta$ as required.

Thus it has been shown that $\mathbf{p}_\beta \subseteq \{\xi \in \mathbf{g} \mid \xi_x \in T_x Y_\beta^{ss}\}$ and that equality holds when $x \in Z_\beta^{ss}$. But the subset of Y_β^{ss} where equality holds is open and is invariant under the action of P_β. So it suffices to show that the only P_β-invariant neighbourhood of Z_β^{ss} in Y_β^{ss} is Y_β^{ss} itself. This follows easily from the fact that if $y \in Y_\beta^{ss}$ then the point $p_\beta(y) \in Z_\beta^{ss}$ lies in the closure of the orbit of x under any one-parameter subgroup of T which is an integer multiple of $\beta \in M(T)$.

This completes the proof of lemma 13.4.

We can now prove the result we were aiming for.

13.5. <u>Theorem.</u> <u>Suppose</u> $X \subseteq \mathbf{P}_n$ <u>is a nonsingular projective variety over</u> k

and G is a reductive subgroup of $GL(n+1)$ defined over k which acts on X. Then the stratification $\{S_\beta | \beta \in \mathbf{B}\}$ of X defined in §12 is smooth. For each $\beta \in \mathbf{B}$ the stratum S_β is isomorphic to $G \times_{P_\beta} Y_\beta^{ss}$ where Y_β^{ss} is a nonsingular locally-closed subvariety of X and P_β is a parabolic subgroup of G. Moreover there is an algebraic locally trivial fibration $p_\beta : Y_\beta^{ss} \to Z_\beta^{ss}$ with affine fibres where Z_β^{ss} consists of the semistable points of a closed nonsingular subvariety of X under the action of a maximal reductive subgroup of P_β.

Proof. By 12.26 for each $\beta \in \mathbf{B}$ the stratum S_β coincides with GY_β^{ss} where Y_β^{ss} is defined as in 12.20. Moreover by 12.23 Y_β^{ss} is invariant under the action of the parabolic subgroup P_β of G defined in 12.11. So there is a morphism

$$\sigma: G \times_{P_\beta} Y_\beta^{ss} \to X$$

whose image is S_β. We shall show using lemma 13.4 that σ is an isomorphism onto its image. The proof is a standard one (cf. e.g. [B] 11.9).

Recall that

$$W_\beta = \{x \in X | x_i = 0 \text{ for } \alpha_i . \beta < q(\beta)\}$$

is invariant under P_β. Consider the morphisms

$$G \times W_\beta \overset{\gamma}{\to} G \times X \overset{\delta}{\to} (G/P_\beta) \times X$$

given by $\gamma(g,x) = (g,gx)$ and $\delta(g,x) = (gP_\beta,x)$. Let

$$M = \delta\gamma(G \times W_\beta)$$

and

$$M' = \delta\gamma(G \times Y_\beta^{ss}).$$

Since W_β is invariant under P_β we have $\delta^{-1}(M) = \{(g,y) | g^{-1}y \in W_\beta\}$ which is closed in $G \times X$ and is isomorphic to $G \times W_\beta$ via the map γ. As δ is a quotient morphism M is therefore closed in $(G/P_\beta) \times X$.

Now GW_β is the image of M under the projection

$$p_X: (G/P_\beta) \times X \to X.$$

Since G/P_β is complete this shows that GW_β is closed (we have already used this). Furthermore

$$G(W_\beta - Y_\beta^{ss}) \subseteq \bigcup_{\beta'>\beta} S_{\beta'}$$

by 12.17 and it follows that $M' = M \cap p_X^{-1}(S_\beta)$ and hence is an open subset of M. We have

$$M' = \{(gP_\beta,y) | g^{-1}y \in Y_\beta^{ss}\}$$

which is isomorphic to $G \times_{P_\beta} Y_\beta^{ss}$ and hence is nonsingular. Moreover by lemma 13.4 the restriction of p_X to M' is a bijection onto S_β. Indeed since G/P_β is complete p_X: $G/P_\beta \times X \to X$ is a closed map, so that p_X: $M' \to S_\beta$ is a homeomorphism because M' is locally closed in $G/P_\beta \times X$. To show that p_X: $M' \to S_\beta$ is an isomorphism it therefore suffices by [Ha] ex I 3.3 and lemma II 7.4 to check that the induced maps of Zariski tangent spaces $(p_X)_*$: $T_mM' \to T_{p_X(m)}S_\beta$ are all injective.

It is only necessary to consider the case when $m = (P_\beta,y)$ for some $y \in Y_\beta^{ss}$. Then an element of T_mM' is of the form $(a + \mathbf{p}_\beta, \xi)$ where $a + \mathbf{p}_\beta \in \mathbf{g}/\mathbf{p}_\beta$, $\xi \in T_yX$ and $-a_y + \xi \in T_yY_\beta^{ss}$. So if $0 = (p_X)_*(a + \mathbf{p}_\beta, \xi) = \xi$ then $a_y \in T_yY_\beta^{ss}$, and hence by lemma 13.4

$a \in \mathbf{p}_\beta$ so that $(a + \mathbf{p}_\beta, \xi)$ is the zero element of $T_m M'$. It follows that $(p_X)_*$ is injective everywhere in M' and hence that $p_X\colon M' \to S_\beta$ is an isomorphism. We conclude that for each $\beta \in \mathbf{B}$ the stratum S_β is nonsingular and isomorphic to $G \times_{P_\beta} Y_\beta^{ss}$.

Thanks to corollary 13.2 the proof of the theorem is now complete.

§14. Hodge numbers

Suppose now that $X \subseteq \mathbf{P}_n$ is a nonsingular complex projective variety acted on linearly by a connected complex reductive group G. Suppose also that the stabiliser in G of every semistable point of X is finite. We have obtained a formula for the Betti numbers of the quotient variety M associated in invariant theory to the action of G on X. In this section we shall see that this formula can be refined to give a formula for the Hodge numbers of M.

We shall use Deligne's extension of Hodge theory which applies to algebraic varieties which are not necessarily compact and nonsingular (see [D1] and [D2]). If Y is a variety which is not nonsingular and projective it may not be possible to decompose $H^n(Y;\mathbf{C})$ as the direct sum of subspaces $H^{p,q}(Y)$ in a way which generalises the classical Hodge decomposition. However Deligne shows that there are two canonical filtrations of $H^n(Y;\mathbf{C})$, the weight filtration

$$\ldots \subseteq W_{k-1} \subseteq W_k \subseteq W_{k+1} \subseteq \ldots$$

which is defined over Q, and the Hodge filtration

$$\ldots \supseteq F_{p-1} \supseteq F_p \supseteq F_{p+1} \supseteq \ldots,$$

giving what Deligne calls a mixed Hodge structure on $H^n(Y)$. One can then define the Hodge numbers $h^{p,q}(H^n(Y))$ of $H^n(Y)$ to be the dimension of appropriate quotients associated to these filtrations ([D1] II 2.3.7). The Hodge numbers satisfy

$$\dim H^n(Y;C) = \sum_{p,q} h^{p,q}(H^n(Y)).$$

If $h^{p,q}(H^n(Y)) \neq 0$ then p and q lie between $\max(0,n-\dim Y)$ and $\min(n,\dim Y)$, and $p+q \leq n$ if Y is projective while $p+q \geq n$ if Y is nonsingular (see [D2] 7 or [D1] III 8.2.4). When Y is nonsingular and projective the Hodge numbers $h^{p,q}(H^n(Y))$ with $p+q = n$ are the same as the classical Hodge numbers $h^{p,q}(Y)$. If $f: Y_1 \to Y_2$ is a morphism of nonsingular quasi-projective varieties then the induced homomorphism $f^*: H^*(Y_2) \to H^*(Y_1)$ is strictly compatible with both the Hodge filtration and the weight filtration (see [D1] II 3.2.11.1).

Suppose now that Y is acted on by a group G. Recall that its equivariant cohomology is defined to be

$$H^*_G(Y;Z) = H^*(Y \times_G EG;Z)$$

where $EG \to BG$ is the universal classifying bundle for G. Although BG is not a finite dimensional manifold there is a natural Hodge structure on its cohomology (see [D1] III 9). Indeed BG may be regarded as the union of finite dimensional varieties M_n such that for any n the inclusion of M_n in BG induces isomorphisms of cohomology in dimensions less than n which preserve the Hodge structure. In the same way $Y \times_G EG$ is the union of finite dimensional varieties whose Hodge structures induce a natural Hodge structure on the cohomology of $Y \times_G EG$. Thus we can define equivariant Hodge numbers

$$h_G^{p,q;n}(Y) = h^{p,q}(H_G^n(Y))$$

for Y.

In particular there are equivariant Hodge numbers for each stratum S_β of the stratification associated in §12 to the action of G on the projective variety X. These strata may be disconnected so it is convenient to refine the stratification as follows. For each integer $m > 0$ let $S_{\beta,m}$ be the union of those components of S_β whose complex codimension in X is $\frac{1}{2}d(\beta,m)$ where

$$d(\beta,m) = m - \dim G + \dim \mathrm{Stab}\beta$$

(cf. §§4 and 8). In §8 we saw that

14.1 $$\dim H_G^n(X;Q) = \sum_{\beta,m} \dim H_G^{n-d(\beta,m)}(S_{\beta,m};Q)$$

for each $n \geq 0$, where the sum is over all $\beta \in \mathbf{B}$ and integers $0 \leq m \leq \dim X$. The argument for this goes as follows. First, because $\{S_{\beta,m} \mid \beta \in \mathbf{B}, 0 \leq m \leq \dim X\}$ is a stratification of X, the elements of the indexing set $\mathbf{B} \times \{0,\ldots,\dim X\}$ can be ordered and renamed as $1,\ldots,M$ for some M in such a way that $S_1 \cup S_2 \cup \ldots \cup S_i$ is open in X for $1 \leq i \leq M$ (see definition 2.11). Let T_i denote this open subset $S_1 \cup S_2 \cup \ldots \cup S_i$ for $1 \leq i \leq M$. Then as each stratum S_i is smooth the Thom isomorphism theorem tells us that

$$H_G^n(T_i,T_{i-1};Q) \cong H_G^{n-2\lambda_i}(S_i;Q)$$

where λ_i is the complex codimension of S_i in X. Thus for each i there is a long exact sequence (the Gysin sequence)

$$\dots \to H_G^{n-2\lambda_i}(S_i;Q) \to H_G^n(T_i;Q) \to H_G^n(T_{i-1};Q) \to H_G^{n+1-2\lambda_i}(S_i;Q) \to \dots$$

In §5 we showed that the stratification is equivariantly perfect over **Q**, which means exactly that each of these long exact sequences splits into short exact sequences

14.2 $$0 \to H_G^{n-2\lambda_i}(S_i;Q) \to H_G^n(T_i;Q) \to H_G^n(T_{i-1};Q) \to 0.$$

Then

14.3 $$\dim H_G^n(T_i;Q) = \dim H_G^n(T_{i-1};Q) + \dim H_G^{n-2\lambda_i}(S_i;Q)$$

for each n, and we obtain the formula 14.1 by using induction on i.

In order to extend the formula 14.1 to Hodge numbers, all we need is the following lemma.

14.4. Lemma. The homomorphism $H_G^n(T_i) \to H_G^n(T_{i-1})$ induced by the inclusion of T_{i-1} in T_i is strictly compatible with the Hodge structures. So is the homomorphism $H_G^{n-2\lambda_i}(S_i) \to H_G^n(T_i)$ except that the Hodge structure of $H_G^{n-2\lambda_i}(S_i)$ must be shifted up by λ_i; that is, the weight filtration $\{W_k\}_{k\in \mathbf{Z}}$ is replaced by $\{W_{k+2\lambda_i}\}_{k\in\mathbf{Z}}$ and the Hodge filtration $\{F_p\}_{p\in\mathbf{Z}}$

by $\{F_{p+\lambda_i}\}_{p\in\mathbb{Z}}$.

Proof. The first statement follows from [D1] II 2.3.5 and 3.2.11.1.

When Y is a nonsingular variety of complex dimension N, Poincaré duality gives us an isomorphism

$$H^n(Y;Q) \cong \mathrm{Hom}(H_c^{2N-n}(Y;Q), H_c^{2N}(Y;Q)) = (H_c^{2N-n}(Y;Q))^*$$

where H_c is cohomology with compact supports and * indicates duality. There is a natural Hodge structure on $(H_c^{2N-n}(Y))^*$ (see [D1] II) and Poincaré duality carries the Hodge structure on $H^n(Y)$ to the natural Hodge structure on $(H_c^{2N-n}(Y))^*$ shifted up by N (see [D2] 8.2). If $i: Y' \to Y$ is the inclusion in Y of a smooth closed subvariety Y' of codimension λ, then the composition

$$H^{n-2\lambda}(Y';Q) \overset{\text{Thom}}{\cong} H^n(Y,Y';Q) \to H^n(Y;Q)$$

is the composition of two Poincaré duality maps with the dual of the map induced by i on cohomology with compact supports:

$$H^{n-2\lambda}(Y';Q) \cong (H_c^{2N-n}(Y';Q))^* \xrightarrow{(i^*)^*} H_c^{2N-n}(Y;Q))^* \cong H^n(Y;Q).$$

Since $(i^*)^*$ is strictly compatible with the Hodge structure we deduce that this composition carries the usual Hodge structure on $H^n(Y)$ to the Hodge structure on $H^{n-2\lambda}(Y')$ shifted up by λ.

The result follows by applying this to finite dimensional approximations to the inclusion of $S_i \times_G EG$ in $T_i \times_G EG$.

It follows from this lemma and the exact sequence 14.2 that

$$h_G^{p,q;n}(T_i) = h_G^{p,q;n}(T_{i-1}) + h_G^{(p,q;n)-\lambda_i}(S_i)$$

where $(p,q;n)-\lambda$ is shorthand for $p-\lambda,q-\lambda;n-2\lambda$. Thus by induction we obtain

14.5 $$h_G^{p,q;n}(X) = h_G^{p,q;n}(X^{ss}) + \sum_{\beta,m} h_G^{(p,q;n)-\frac{1}{2}d(\beta,m)}(S_{\beta,m})$$

where the sum is over all nonzero $\beta \in \mathbf{B}$ and integers $0 \leq m \leq \dim X$.

By theorem 13.7 for each $\beta \in \mathbf{B}$ we have

$$S_\beta \cong G \times_{P_\beta} Y_\beta^{ss},$$

and there is a locally trivial fibration

$$p_\beta : Y_\beta^{ss} \to Z_\beta^{ss}$$

with contractible fibre which respects the action of $\mathrm{Stab}\beta$. Since $\mathrm{Stab}\beta$ is homotopically equivalent to P_β it follows that

$$H_G^*(S_\beta;Q) \cong H_{P_\beta}^*(Y_\beta^{ss};Q) \cong H_{\mathrm{Stab}\beta}^*(Z_\beta^{ss};Q),$$

and it is easily checked that these are isomorphisms of the Hodge structures. By looking at components we also get

$$H_G^*(S_{\beta,m};Q) \cong H_{\mathrm{Stab}\beta}^*(Z_{\beta,m}^{ss};Q)$$

for each β and m, where $Z_{\beta,m}^{ss}$ is the set of semistable points of a nonsingular subvariety $Z_{\beta,m}$ of X under a suitable linearisation of the action of $\mathrm{Stab}\beta$. Hence

14.6 $$h_G^{p,q;n}(S_{\beta,m}) = h_{\mathrm{Stab}\beta}^{p,q;n}(Z_{\beta,m}^{ss})$$

for each p, q and n. Therefore

14.7 $$h_G^{p,q;n}(X^{ss}) = h_G^{p,q;n}(X) - \sum_{\beta,m} h_{\mathrm{Stab}\beta}^{(p,q;n)-\frac{1}{2}d(\beta,m)}(Z_{\beta,m}^{ss}).$$

This gives us an inductive formula for the equivariant Hodge numbers of X^{ss} in terms of those of X itself and of the semistable strata of smaller varieties acted on by reductive groups.

We also know that the fibration $X \times_G EG \to BG$ with fibre X is cohomologically trivial over $\mathbf{Q}$ (see theorem 5.4), so that

14.8 $$H_G^*(X;\mathbf{Q}) \cong H^*(X;\mathbf{Q}) \otimes H^*(BG;\mathbf{Q}).$$

This isomorphism is an isomorphism of Hodge structures ([D1] III 8.2.10).

Using 14.7 and 14.8 an explicit formula can be derived for the equivariant Hodge numbers of the semistable stratum X^{ss}. This formula involves the Hodge numbers of X and certain nonsingular subvarieties of X, and also the Hodge numbers of the classifying space of G and various reductive subgroups of G (cf. §5).

By assumption the stabiliser in G of every $x \in X^{ss}$ is finite. This implies that the quotient variety M coincides with the topological quotient X^{ss}/G (see §8). Moreover the obvious map $X^{ss} \times_G EG \to X^{ss}/G$ induces an isomorphism

$$H^*(X^{ss}/G;Q) \to H_G^*(X^{ss};Q)$$

which is strictly compatible with the Hodge structures and hence is an isomorphism of Hodge structures. Thus we obtain a formula for calculating the Hodge numbers $h^{p,q;n}(M)$ of the quotient $M = X^{ss}/G$, which are the classical Hodge numbers $h^{p,q}(M)$ when M is smooth.

Note that since $h_G^{p,q;n}(X)$ is nonzero only when $p+q=n$ the same is true by induction of $h_G^{p,q;n}(X^{ss})$ and each $h_G^{p,q;n}(S_{\beta,m})$, and hence also of $h^{p,q;n}(M)$ when the stabiliser of every $x \in X^{ss}$ is finite. This last fact could of course also be deduced directly from [D1] III 8.2.4 and the fact that X^{ss}/G is a compact rational homology manifold.

Finally note that 14.2 shows that the map $H_G^n(X;Q) \to H_G^n(X^{ss};Q)$ induced by the inclusion of X^{ss} in X is surjective, since it is the composition of the surjective maps $H_G^n(T_i;Q) \to H_G^n(T_{i-1};Q)$ for $1 \leq i \leq M$. Thus we have a surjective homomorphism

14.9 $$H^*(X;Q) \otimes H^*(BG;Q) \to H^*(M;Q)$$

which is strictly compatible with the Hodge structures. In particular if $h^{p,q}(X) = 0$ for $p \neq q$ then the same is true for M, because by [D1] III 9.1.1 only the even Betti numbers of BG are nonzero and $H^{2n}(BG;\mathbf{C})$ is purely of type (n,n) for every n.

§15. Calculating cohomology by counting points

Again let M be the projective quotient variety associated to the linear action of a complex reductive group G on a nonsingular complex projective variety X. When the action of G on the semistable stratum X^{ss} is free, there is an alternative method of deriving the formulae already obtained for the Betti numbers of M which uses the Weil conjectures. These conjectures, which were verified by Deligne, enable one to calculate the Betti numbers of a nonsingular complex projective variety by counting the number of points in associated varieties defined over finite fields $\mathbf{F}_q$. In our case we can count points by using the stratifications defined in §12 of varieties over the algebraic closures $\overline{\mathbf{F}}_q$. This idea was suggested by work of Harder and Narasimhan, who used the Weil conjectures to calculate Betti numbers of moduli spaces of bundles over Riemann surfaces (see [H & N]). Their formulae were subsequently rederived in the paper of Atiyah and Bott which motivated Part I.

The idea of the alternative method is explained in this section, but the arguments are not given in detail because nothing new is being proved. Unless the Weil conjectures can be extended in an appropriate way to projective varieties which are locally the quotients of nonsingular varieties by finite groups, the same method will not work in all cases where the stabiliser of each semistable point is finite. It is necessary that the action of G or at least some quotient of G on X^{ss} be free.

First let us summarise what we shall need of the Weil conjectures (for more details see [B-B&S] §2).

Let Y be a nonsingular complex projective variety. Then Y is defined over a finitely-generated subring R of $\mathbb{C}$, so that there is an R-scheme Y_R such that $Y = Y_R \times_R \operatorname{Spec} \mathbb{C}$. Let π be a maximal ideal of R. Then R/π is a finite field with q elements for some prime power q. Let

15.1 $$Y_\pi = Y_R \times_R \operatorname{Spec} R/\pi$$

be the reduction of Y modulo π. For most choices of π, if ℓ is any prime number different from the characteristic of R/π then the ℓ-adic Betti numbers of Y_π and Y are equal. But the ℓ-adic Betti numbers of Y are the same as the ordinary Betti numbers of Y regarded as a complex manifold, by the comparison theorem of ℓ-adic cohomology.

Provided that the characteristic of R/π is not one of finitely many "bad" primes, Y_π is a nonsingular projective variety over the finite field with q elements. Then the Weil conjectures enable us to calculate its ℓ-adic Betti numbers. In fact there exist complex numbers $\alpha_1,\dots,\alpha_r$, $\beta_1,\dots,\beta_s$ such that for any integer $n \geq 1$ the number of points of Y_π defined over the finite field with q^n elements is

15.2 $$\sum_i (\alpha_i)^n - \sum_j (\beta_j)^n .$$

We may assume that $\alpha_i \neq \beta_j$ for every i and j. Then the absolute value of each α_i is of the form $q^{n(i)}$ and the absolute value of each β_j is of the form $q^{n(j)+\frac{1}{2}}$ where $n(i)$ and $n(j)$ are non-negative integers. Moreover the $(2k)$th ℓ-adic Betti number of Y_π is equal to the number of α_i's with absolute value q^k, and its $(2k+1)$st ℓ-adic Betti number is the number of β_j's with absolute value $q^{k+\frac{1}{2}}$.

We shall use the Weil conjectures in a slightly different but equivalent form.

15.3. <u>Definition.</u> For $r \geq 1$ let $N_r(Y)$ be the number of points of Y_π which are defined over the field of q^r elements. If n is the dimension of Y let $\tilde{N}_r(Y) = q^{-rn} N_r(Y)$.

15.4. It follows easily from Poincaré duality and the Weil conjectures as stated above that we can write the series

$$\exp\left(\sum_{r \geq 1} \tilde{N}_r(Y) t^r / r\right) \in Q[[t]]$$

in the form

$$Q_1(t) Q_3(t) \ldots Q_{2n-1}(t) / Q_0(t) Q_2(t) \ldots Q_{2n}(t)$$

where

$$Q_i(t) = \prod_j (1 - \gamma_{ij}t)$$

for complex numbers γ_{ij} satisfying

$$|\gamma_{ij}| = q^{-i/2}$$

and where $\deg Q_i$ is the ith Betti number of Y.

We shall use 15.4 to calculate the rational Poincaré polynomial of the quotient variety M associated to the action of G on X. (It seems to be natural to use this dual form of the Weil conjectures here. This is what Atiyah and Bott do when comparing their methods with those of [H & N]. Using the ordinary form corresponds to using cohomology with compact supports, and it is difficult to make sense of this for the infinite-dimensional manifolds in [A & B]).

For simplicity suppose that G is a subgroup of $GL(n+1)$. We assume that G acts freely on X^{ss}. The argument we shall use runs as follows.

We may assume throughout that the action of G on X is defined over R and that all the (finitely many) quasi-projective nonsingular subvarieties of X and subgroups of G which we shall need to consider are also defined over R and have nonsingular reduction modulo π. We may also assume that their dimensions are unaltered by reduction modulo π. Moreover the Weil conjectures still hold if q is replaced by some power q^s. Hence we may assume that all subvarieties of X_π and subgroups of G_π under consideration are defined over the field $\mathbf{F}_q$ with q elements.

We shall find that the stratification of X_π induced by the action of G_π is the reduction modulo π of the stratification of X induced by the action of G, and hence using the results of §13 that

$$15.5 \qquad \tilde{N}_r(X^{ss}) = \tilde{N}_r(X) - \sum_{\beta,m} q^{-\frac{1}{2} rd(\beta,m)} \tilde{N}_r(Z^{ss}_{\beta,m}) \tilde{N}_r(G/P_\beta)$$

where the sum is over all nonzero $\beta \in \mathbf{B}$ and integers $0 \leq m \leq \dim X$. This gives us an inductive formula for $\tilde{N}_r(X^{ss})/\tilde{N}_r(G)$ which is analogous to the formula for the Poincaré series $P^G_t(X^{ss})$ obtained in Part I. From it an explicit formula can be derived for $\tilde{N}_r(M)$ by the arguments used in §5. This formula is such that if q is replaced by t^{-2} and $\tilde{N}_r(Y)$ by $P_t(Y)$ for each projective variety Y which appears in it, then the result is the formula for $P_t(M)$ already derived. It then remains only to justify this substitution.

Let us now examine the details of this argument more closely.

Let T be a maximal torus of G defined over R and let $B \supseteq T$ be a Borel subgroup also defined over R. By extending R if necessary we may assume that T acts diagonally on R^{n+1}. It follows from our assumptions that the group G_π is reductive and has T_π as a maximal torus and B_π as a Borel subgroup.

Theorem 12.26 can be applied to the action of G_π on X_π and to that of G on X to obtain stratifications of X_π and X. It is necessary to investigate the relationship between these stratifications. First we must check that they can be indexed by the same set **B**. Recall that the indexing set for the stratification of X is a finite subset of the **Q**-vector space $M(T) = Y(T) \otimes_{\mathbf{Z}} \mathbf{Q}$ where $Y(T)$ is the free **Z**-module consisting of all one-parameter subgroups of the maximal torus T. Since T_π has the same rank as T there is a natural identification of $M(T)$ with $M(T_\pi)$. The Weyl group actions coincide under these identifications, and so do the weights $\alpha_0,\ldots,\alpha_n$ of the representations of T and T_π which define their actions on X and X_π. We may assume that the norms chosen on $M(T)$ and $M(T_\pi)$ also coincide. Hence the stratifications of X and X_π may be indexed by the same set **B** (see 12.8).

Let $\{S_\beta \mid \beta \in \mathbf{B}\}$ be the stratification of X and let $\{S_{\beta,\pi} \mid \beta \in \mathbf{B}\}$ be the stratification of X_π. Under the assumptions already made the following lemma follows without difficulty from the definitions of §12.

15.6. Lemma. The stratification $\{S_\beta \mid \beta \in \mathbf{B}\}$ is defined over R and

$$(S_\beta)_\pi = S_{\beta,\pi} \quad \text{for each } \beta \in \mathbf{B}.$$

Moreover $(Y_\beta^{ss})_\pi$, $(Z_\beta^{ss})_\pi$ and $(P_\beta)_\pi$ coincide with the subvarieties of X_π and parabolic subgroup of G_π defined in the corresponding way for the action of G_π on X_π. Finally the quotient variety $M = X^{ss}/G$ satisfies

$$\tilde{N}_r(M) = \tilde{N}_r(X^{ss})\, \tilde{N}_r(G)^{-1}$$

for every $r \geq 1$.

In order to apply 15.4 we need to calculate $\tilde{N}_r(M)$ for each $r \geq 1$. The last lemma suggests that we should investigate $\tilde{N}_r(X^{ss})$. It also tells us for each $\beta \in B$ that $N_r(S_\beta)$ is the number of points in the stratum $S_{\beta,\pi}$ of X_π which are defined over the field of q^r elements, and so

15.7 $$N_r(X^{ss}) = N_r(X) - \sum_{\beta \neq 0} N_r(S_\beta).$$

Moreover

$$S_{\beta,\pi} \cong G_\pi \times_{(P_\beta)_\pi} (Y_\beta^{ss})_\pi$$

by the lemma together with theorem 2.26, and so

15.8 $$N_r(S_\beta) = N_r(Y_\beta^{ss}) N_r(G/P_\beta)$$

for each $\beta \in B$. As in §4 we can decompose Y_β^{ss} into a disjoint union of open subsets $\{Y_{\beta,m}^{ss} | 0 \leq m \leq \dim X\}$ such that each component of $Y_{\beta,m}^{ss}$ has real codimension m in X. Then S_β is the disjoint union of open subsets $GY_{\beta,m}^{ss}$ which have complex codimension $\frac{1}{2}d(\beta,m) = \frac{1}{2}m - \dim(G/P_\beta)$ in X. There is also a locally trivial fibration

$$p_\beta : (Y_{\beta,m}^{ss})_\pi \to (Z_{\beta,m}^{ss})_\pi$$

such that each fibre is an affine space (see 13.2), from which it follows that

$$\tilde{N}_r(Y^{ss}_{\beta,m}) = \tilde{N}_r(Z^{ss}_{\beta,m})$$

for each $r \geq 1$. So by 15.7 and 15.8 we have

$$\tilde{N}_r(X^{ss}) = \tilde{N}_r(X) - \sum_{\beta,m} q^{-\frac{1}{2}rd(\beta,m)} \tilde{N}_r(Z^{ss}_{\beta,m}) \tilde{N}_r(G/P_\beta) \tag{15.9}$$

for each $r \geq 1$, where the sum is over nonzero $\beta \in \mathbf{B}$ and integers $0 \leq m \leq \dim X$.

Next we consider $\tilde{N}_r(G/P_\beta)$. As at 6.9 we have $P_\beta = B\,\mathrm{Stab}\beta$ where B is the Borel subgroup of G and $\mathrm{Stab}\beta$ is the stabiliser of β under the adjoint action of G. Since $\mathrm{Stab}\beta$ contains the maximal torus T it follows that $P_\beta = B_u\mathrm{Stab}\beta$ where B_u is the unipotent part of B (see [B] 10.6 (4)).

15.10. <u>Lemma.</u> <u>If</u> H <u>is a unipotent subgroup of</u> G <u>defined over</u> R <u>such that</u> $\dim H_\pi = \dim H$, <u>then</u> $N_r(H) = q^{r\dim H}$ <u>and hence</u> $\tilde{N}_r(H) = 1$ <u>for all</u> $r \geq 1$.

<u>Proof.</u> The remark at the end of [B] 14.4 shows that H_π is isomorphic as a variety over F_q to an affine space. The result follows.

Under our assumptions this lemma applies to the unipotent subgroups B_u and $B_u \cap \mathrm{Stab}\beta$ of G. Hence

15.11 $$\tilde{N}_r(P_\beta) = \tilde{N}_r(\mathrm{Stab}\beta).$$

From this together with 5.9 it follows that

15.12 $$\tilde{N}_r(X)\tilde{N}_r(G)^{-1} = \tilde{N}_r(X^{ss})\tilde{N}_r(G)^{-1}$$

$$+ \sum_{\beta,m} q^{-\frac{1}{2}rd(\beta,m)}\tilde{N}_r(Z_{\beta,m}{}^{ss})\tilde{N}_r(\mathrm{Stab}\beta)^{-1}$$

for all $r \geq 1$, where the sum is over all nonzero $\beta \in \mathbf{B}$ and $0 \leq m \leq \dim X$. This is an inductive formula for $\tilde{N}_r(X^{ss})\tilde{N}_r(G)^{-1}$ (which coincides with $\tilde{N}_r(M)$ under the assumption that G acts freely on X^{ss} by lemma 15.6). By the argument used in §5 we can derive from it the following explicit formula.

15.13 $$\tilde{N}_r(M) = \tilde{N}_r(X)\tilde{N}_r(G)^{-1}$$

$$+ \sum_{\underline{\beta},m} (-1)^{q(\underline{\beta})} q^{-\frac{1}{2}rd(\underline{\beta},m)}\tilde{N}_r(Z_{\underline{\beta},m})\tilde{N}_r(\mathrm{Stab}\underline{\beta})^{-1}$$

for each $r \geq 1$, where the sum is over all integers $0 \leq m \leq \dim X$ and β-sequences $\underline{\beta}$ defined as in §5. If $\underline{\beta} = (\beta_1,\ldots,\beta_q)$ is a β-sequence then $q(\underline{\beta}) = q$ is the length of $\underline{\beta}$. Each $Z_{\underline{\beta},m}$ is a nonsingular closed subvariety of X and $\mathrm{Stab}\underline{\beta} = \bigcap_j \mathrm{Stab}\beta_j$ is a reductive subgroup of G.

From 15.4 we know that the Poincaré polynomial $P_t(Y)$ of any nonsingular projective variety Y can be calculated from the numbers $\tilde{N}_r(Y)$. Our aim is to apply this to the quotient variety M and use the formula 15.13 to obtain an expression for $P_t(M)$. However the groups which appear in 15.13 are not projective varieties, so we need to modify the formula a little as follows.

The Borel subgroup B is the product of its unipotent part B_u and the maximal torus T so lemma 15.10 implies that

15.14 $$\tilde{N}_r(G) = \tilde{N}_r(G/B)\tilde{N}_r(T) = \tilde{N}_r(G/B)(1-q^{-r})^{\dim T}$$

for $r \geq 1$.

If we apply this to each of the subgroups $\mathrm{Stab}\underline{\beta}$ of G and substitute in 15.13, we obtain an expression for $\tilde{N}_r(M)$ as a rational function of q and the numbers $\tilde{N}_r(Y)$ for certain nonsingular projective varieties Y. The varieties involved are X and its subvarieties $Z_{\underline{\beta},m}$ together with the flag manifolds of G and its subgroups $\mathrm{Stab}\underline{\beta}$. This gives us a formula for the Poincaré polynomial $P_t(M)$ of the quotient M because of the following lemma.

15.15. <u>Lemma.</u> <u>Suppose that</u> $Y_1,\ldots,Y_k$ <u>are smooth complex projective varieties defined over</u> R <u>whose reductions modulo</u> π <u>are also smooth. Suppose that</u> f <u>is a rational function of</u> $s+1$ <u>variables with integer coefficients such that</u>

$$f(q^{-r}, \tilde{N}_r(Y_1), \ldots, \tilde{N}_r(Y_k)) = 0$$

for all $r \geq 1$. Then

$$f(t^2, P_t(Y_1), \ldots, P_t(Y_k)) = 0.$$

Proof. Call a sequence $N = (n_r)_{r \geq 1}$ of integers a Weil sequence if there exist finitely many polynomials $Q_i(t)$ of the form

$$Q_i(t) = \prod_j (1 - \gamma_{ij} t),$$

with

$$|\gamma_{ij}| = q^{-i/2}$$

for each i and j, and such that

$$\exp\left(\sum_{r \geq 1} n_r t^r / r\right) = Q_1(t) \ldots Q_{2n-1}(t) / Q_0(t) \ldots Q_{2n}(t)$$

for some $n \geq 0$. These conditions determine each nontrivial Q_i uniquely, so we may define a polynomial $P_t(N)$ by

$$P_t(N) = \sum_{i \geq 0} (\deg Q_i)\, t^i .$$

It is easy to check that if $N = (n_r)_{r \geq 1}$ and $M = (m_r)_{r \geq 1}$ are Weil sequences then so are $NM = (n_r m_r)_{r \geq 1}$, $N+M = (n_r + m_r)_{r \geq 1}$ and $q^{-1}N = (q^{-r} n_r)_{r \geq 1}$, and that $P_t(NM) = P_t(N) P_t(M)$, $P_t(N+M) = P_t(N) + P_t(M)$ and $P_t(q^{-1}N) = t^2 P_t(N)$.

For each positive integer $j \leq k$ let N_j be the sequence $(\tilde{N}_r(Y_j))_{r \geq 1}$. It follows immediately from remark 5.4 that each of the sequences N_j is a Weil sequence and that the polynomial $P_t(N_j)$ coincides with the Poincaré polynomial $P_t(Y_j)$ of Y_j.

To prove the lemma it is enough to consider the case when $f \in \mathbf{Z}[X_0,\ldots,X_n]$. We can write such an f as

$$f = g - h$$

where g and h are sums of monomials with positive integer coefficients. Since $N_1,\ldots,N_k$ are Weil sequences so are the sequences whose rth terms are

$$g(q^{-r},\tilde{N}_r(Y_1),\ldots,\tilde{N}_r(Y_k))$$

and

$$h(q^{-r},\tilde{N}_r(Y_1),\ldots,\tilde{N}_r(Y_k)),$$

and their corresponding polynomials are

$$g(t^2,P_t(Y_1),\ldots,P_t(Y_k))$$

and

$$h(t^2,P_t(Y_1),\ldots,P_t(Y_k)).$$

But by assumption these sequences are equal, and hence so are the corresponding polynomials. The result follows.

This lemma may be applied to the equation obtained from 15.13 by using 15.14 to substitute for $\tilde{N}_r(G)$ and for each $\tilde{N}_r(\text{Stab}\underline{\beta})$. This gives us the following formula for the Betti numbers of the quotient M.

15.16 $$P_t(M) = (1-t^2)^{-\dim T}\{P_t(X)\,P_t(G/B)^{-1} + \sum_{\underline{\beta},m}(-1)^{q(\underline{\beta})}t^{d(\beta,m)}P_t(Z_{\underline{\beta},m})\,P_t(\mathrm{Stab}\underline{\beta}/B\cap\mathrm{Stab}\underline{\beta})^{-1}\}.$$

As before, let BG be the universal classifying space for the group G. There is a fibration $BG \to BT$ which has fibre G/B and is cohomologically trivial. Thus

$$P_t(BG) = P_t(BT)\,P_t(G/B)^{-1} = (1-t^2)^{-\dim T}\,P_t(G/B)^{-1}.$$

By applying this fact to all the reductive subgroups $\mathrm{Stab}\underline{\beta}$ of G, we find that the formula 15.16 for $P_t(M)$ coincides with the formula derived in Part I.

§16. Examples

In this section the stratifications induced by some particular group actions will be described and the Betti numbers of their quotients will be calculated.

We shall start by reviewing the diagonal action of SL(2) on a power $(\mathbf{P}_1)^n$ of the complex projective line. This was used as an example throughout Part I. When SL(2) acts on $\mathbf{P}_n$ identified with the space of binary forms of degree n very similar results hold. Then we shall consider the action of SL(m) on a product of the form

$$X = \prod_j G(\ell_j, m)$$

where $G(\ell, m)$ is the Grassmannian of ℓ-dimensional subspaces of $\mathbf{C}^m$. The subvarieties Z_β which appear in the inductive formula for $P_t^{SL(m)}(X^{ss})$ are all products of varieties of the same form as X but with smaller values of m. Thus although the calculation of $P_t^G(X^{ss})$ for large m would be extremely lengthy by hand, it could be carried out by a computer. We do some explicit calculations for the special case of products $(\mathbf{P}_2)^n$ of the projective plane. These examples are more intricate than $(\mathbf{P}_1)^n$ and are more typical of the general case.

16.1. Ordered points on the projective line (cf. [N] 4 §5)

For fixed $n \geq 1$ consider the diagonal action of the special linear group $SL(2)$ on $(\mathbf{P}_1)^n$. This is linear with respect to the Segre embedding; the corresponding representation of $SL(2)$ is the nth tensor power of its standard representation on $\mathbf{C}^2$. Let T_c be the complex maximal torus consisting of all diagonal elements of $SL(2)$ and let α be the one-parameter subgroup of T_c given by

$$z \to \begin{bmatrix} z & 0 \\ 0 & z^{-1} \end{bmatrix}.$$

The weights of the representation with respect to this torus are of the form

$$r\alpha - (n-r)\alpha$$

where r is any integer such that $0 \leq r \leq n$. If we choose the positive Weyl chamber to contain α then it follows that the indexing set for the stratification is

$$\mathbf{B} = \{ (2r-n)\alpha \mid n \geq r > n/2\} \cup \{0\} .$$

Suppose $\beta = (2r-n)\alpha$ where $r > n/2$. Then it is easy to check from definition 12.18 that a sequence in $(\mathbf{P}_1)^n$ lies in Z_β if and only if it contains r copies of 0 and $n-r$ copies of ∞. Also Y_β consists of all sequences containing precisely r copies of 0.

It follows from definition 12.20 that $Z_\beta^{ss} = Z_\beta$ and hence that $Y_\beta^{ss} = Y_\beta$. Since the stratum S_β indexed by β is GY_β^{ss} (see 2.26) it follows that S_β consists of all sequences $(x_1,\ldots,x_n)$ such that r but no more of the points $x_1,\ldots,x_n$ coincide. Thus S_β has $\binom{n}{r}$ components, each of which has complex codimension $r-1$.

Therefore the semistable elements of $(\mathbf{P}_1)^n$ are those which contain no point of $\mathbf{P}_1$ with multiplicity strictly greater than $n/2$ (cf. [N] 4 §5). If $x \in (\mathbf{P}_1)^n$ is not semistable, the stratum to which x belongs is determined by the multiplicity of the unique point of $\mathbf{P}_1$ which occurs as a component of x strictly more than $n/2$ times.

SL(2) is the complexification of the compact group SU(2) which preserves the standard Kåhler structure on $(\mathbf{P}_1)^n$. Since SU(2) is semisimple there is a unique moment map $\mu\colon (\mathbf{P}_1)^n \to \mathbf{su}(2)$. The adjoint action of SU(2) on its Lie algebra $\mathbf{su}(2) \cong \mathbf{R}^3$ is via the double cover $\theta\colon$ SU(2) $\to$ SO(3). Use the standard inner product on $\mathbf{R}^3$ to identify $\mathbf{su}(2)$ with its dual. The complex projective line $\mathbf{P}_1$ may be identified with the unit sphere in $\mathbf{R}^3$, which is an orbit of the adjoint representation of SU(2). By [Ar] p. 322 the moment map for the action of SU(2) on $\mathbf{P}_1$ is then the inclusion $\mathbf{P}_1 \to \mathbf{R}^3$. It follows easily from this or from the formula 2.7 that the moment map

$$\mu\colon (\mathbf{P}_1)^n \to \mathbf{R}^3$$

is given by

$$\mu(x_1,\ldots,x_n) = x_1 + \ldots + x_n .$$

So

$$f(x_1,\dots,x_n) = \|x_1 + \dots + x_n\|^2$$

where $\| \ \|$ is the standard norm on $\mathbf{R}^3$. A point $(x_1,\dots,x_n)$ is critical for f if either $f(x_1,\dots,x_n) = 0$ or each x_i is one of a fixed pair of antipodal points of $\mathbf{P}_1$.

It is intuitively reasonable that the Morse stratification of this function should coincide with the stratification $\{S_\beta \mid \beta \in \mathbf{B}\}$ already described. For by symmetry if two components x_i and x_j of $x = (x_1,\dots,x_n)$ agree then these components will remain the same on the path of steepest descent for f from x. On the other hand it is possible to move a configuration of n points into a balanced position (i.e. a position with centre of gravity at the origin) without splitting up points which coincide if and only if no point has multiplicity strictly more than $n/2$.

Note that the stratification for the action of $GL(2)$ is the same as that for $SL(2)$ although labelled differently. This is because $GL(2)$ is the quotient by a finite subgroup of the product of $SL(2)$ with a central one-parameter subgroup which acts trivially on $\mathbf{P}_1$.

The stabiliser in $PGL(2)$ of a point $x \in (\mathbf{P}_1)^n$ is non-trivial precisely when at most two distinct points of $\mathbf{P}_1$ occur as components of x. So if n is odd $PGL(2)$ acts freely on the semistable points of $(\mathbf{P}_1)^n$. Then as $SL(2)$ is a finite cover of $PGL(2)$ we can use theorem 8.12 to calculate the Betti numbers and Hodge numbers of the quotient variety M as follows.

Since the rank of SU(2) is 1 each β-sequence has length 1 (see definition 5.11) and so is just a nonzero element of **B**. Thus by 8.10 and 5.17

$$P_t(M) = P_t((\mathbf{P}_1)^n)P_t(BSU(2)) - \sum_{n/2<r\leq n} \binom{n}{r} t^{2(r-1)}P_t(BS^1)$$

$$= (1+t^2)^n(1-t^4)^{-1} - \sum_{n/2<r\leq n} \binom{n}{r} t^{2(r-1)}(1-t^2)^{-1} \qquad (^*)$$

$$= 1 + nt^2 + \dots + \{1 + (n-1) + \binom{n-1}{2} + \dots$$

$$+ \binom{n-1}{\min(j,n-3-j)}\} t^{2j} + \dots + t^{2n-6} .$$

This obeys Poincaré duality as expected. Note that the equivariant cohomology of the semistable stratum of $(\mathbf{P}_1)^n$ is given by the series $(^*)$ for any n, even or odd. However this is not a polynomial when n is even.

When n is odd it is also possible to obtain the Hodge numbers of the quotient M. Indeed 14.9 shows that

$$h^{p,q}(M) = 0 \quad \text{for} \quad p \neq q,$$

and

$$h^{p,p}(M) = 1 + (n-1) + \dots + \binom{n-1}{\min(p,n-3-p)}$$

for each p.

16.2. Binary forms (cf. [M] 4 §1 and [N] 4 §1, §3)

An example which is similar to 16.1 is the action of $SL(2)$ on the projective space $\mathbf{P}_n$ identified with the nth symmetric product of $\mathbf{P}_1$.

The maximal torus T_c acts on $\mathbf{P}_n$ via the homomorphism

$$\operatorname{diag}(z,z^{-1}) \to \operatorname{diag}(z^n,z^{n-2},\ldots,z^{-n}),$$

so that as in 16.1

$$\mathbf{B} = \{ (2r-n)\alpha \mid n \geq r > n/2\} \cup \{0\}.$$

If $\beta = (2r-n)\alpha \in \mathbf{B}$ then $Z_\beta^{ss} = Z_\beta$ consists of the single configuration in which 0 has multiplicity r and ∞ has multiplicity $n-r$. The stratum S_β consists of those configurations with a point of multiplicity precisely r and has codimension $r-1$ in $\mathbf{P}_n$.

Thus the stratifications of $(\mathbf{P}_1)^n$ and $\mathbf{P}_n$ correspond under the quotient map $h: (\mathbf{P}_1)^n \to \mathbf{P}_n$. However the moment maps do not correspond. This reflects the fact that the symplectic structure is not preserved by h. The Kåhler form on $\mathbf{P}_n$ pulls back via h to a form on $(\mathbf{P}_1)^n$ which is symplectic except that it degenerates along a subset of positive codimension. In fact such forms give moment maps in the same way as nondegenerate ones. Thus we have two different moment maps on $(\mathbf{P}_1)^n$ which induce the same stratification.

When n is odd the stabilisers of all semistable points are finite, so there is a (singular) projective quotient $M = \mathbf{P}_n^{ss}/SL(2)$ such that

$$P_t(M) = (1 + t^2 + \ldots + t^{2n})(1 - t^4)^{-1} - \sum_{n \geq r > n/2} t^{2(r-1)}(1 - t^2)^{-1}$$

$$= (1-t^2)^{-1}(1+t^4+\dots+t^{n-1}-t^{n+1}-\dots-t^{2(n-1)})$$

$$= 1+t^2+2t^4+2t^6+3t^8+\dots+[1+\tfrac{1}{2}\min(j,n-3-j)]t^{2j}+\dots+t^{2n-6}.$$

16.3. Products of Grassmannians

This example is a generalisation of 16.1. If V is a complex vector space let $G(\ell,V)$ be the Grassmannian of ℓ-dimensional linear subspaces of V, or equivalently of $(\ell-1)$-dimensional linear subspaces of the projective space $\mathbf{P}(V)$. We can embed $G(\ell,V)$ in $\mathbf{P}(\Lambda^{\ell}V)$ by using Plücker coordinates. Thus any product of the form

$$X = G(\ell_1,\mathbf{C}^m) \times \dots \times G(\ell_r,\mathbf{C}^m)$$

can be embedded as a subvariety of the projective space $\mathbf{P}(\bigotimes_{1\le j\le r} \Lambda^{\ell_j}\mathbf{C}^m)$ on which $GL(m)$ acts linearly.

Since the central one-parameter subgroup of $GL(m)$ acts trivially on $\bigotimes_{1\le j\le r} \Lambda^{\ell_j}\mathbf{C}^m$ the stratification of X arising from this action of $GL(m)$ coincides with the $SL(m)$ stratification, except that a stratum labelled by β for $GL(m)$ is labelled in the $SL(m)$ stratification by the projection

$$16.4 \qquad \beta - (n+1)^{-1}\Big(\sum_{1\le j\le r} \ell_j\Big)(1,\dots,1)$$

of β onto the Lie algebra of $SU(m)$.

By [M] 5.3 or [N] 4.17 a sequence of subspaces $(L_1,\dots,L_r) \in X$ is semistable for $SL(M)$ if and only if

16.5 $$\sum_{1\leq j\leq r} (\dim L_j \cap M)/(\dim M) \leq \sum_{1\leq j\leq r} \ell_j/m$$

for every proper subspace $M \subseteq \mathbf{C}^m$. The stabiliser of $(L_1,\ldots,L_r)$ is finite if strict inequality always holds. Therefore if

$$\sum_{1\leq j\leq r} \ell_j$$

is coprime to m, every semistable point of X has finite stabiliser and so theorem 8.12 will give us a formula for the Betti numbers of the quotient variety.

Suppose that $(L_1,\ldots,L_r)$ is not semistable. Let $\mathbf{M}$ be the set of proper subspaces M of $\mathbf{C}^m$ such that the ratio

$$\sum_{1\leq j\leq r} (\dim L_j \cap M) / (\dim M)$$

is maximal. Then by 16.5 for each $M \in \mathbf{M}$ the sequence $(L_1 \cap M,\ldots,L_r \cap M)$ is semistable in

$$\prod_{1\leq j\leq r} G(\dim L_j \cap M, M).$$

Let M_1 be a maximal element of $\mathbf{M}$. If $M \in \mathbf{M}$ it is easy to check that $M + M_1$ also lies in $\mathbf{M}$, and hence $M \subseteq M_1$ by the maximality of M_1. In particular M_1 is uniquely determined.

By induction we find that any $(L_1,\ldots,L_r) \in X$ determines a unique sequence

$$0 = M_0 \subsetneq M_1 \subsetneq \ldots \subsetneq M_s = C^m$$

of subspaces of C^m satisfying the following conditions.

16.7(a) The sequence $(L_{i1},\ldots,L_{ir})$ is semistable in

$$\prod_{1\le j\le r} G(\ell_{ij}, M_i/M_{i-1})$$

for $1 \le i \le s$, where

$$L_{ij} = (L_j \cap M_i + M_{i-1})/M_{i-1}$$

is the image of L_j in M_i/M_{i-1}, and

$$\ell_{ij} = \dim L_{ij} \text{ .}$$

(b) Each M_i is maximal among subspaces with property (a).

(c) $\sum_{1\le j\le r} (\dim L_j \cap M_i)/(\dim M_i) > \sum_{1\le j\le r} (\dim L_j \cap M_{i-1})/(\dim M_{i-1})$

for $1 \le i \le s$, or equivalently

$$k_1/m_1 > k_2/m_2 > \ldots > k_s/m_s$$

where $k_i = \sum_{1\le j\le r} \ell_{ij}$ and $m_i = \dim M_i/M_{i-1}$.

16.8. <u>Remark.</u> The equivalence in (c) comes from the fact that if $a,b,c,d > 0$ then $a/b < c/d$ if and only if $(a + c)/(b + d) < c/d$.

Let T_c be the complex maximal torus of $GL(m)$ consisting of the diagonal matrices, and let $T = T_c \cap U(m)$. Denote by $\mathbf{t}_+$ the standard positive Weyl chamber in the Lie algebra of T.

16.9. Proposition. Suppose $(L_1,\ldots,L_r) \in X$. Let

$$0 = M_0 \subsetneq M_1 \subsetneq \ldots \subsetneq M_s = \mathbf{C}^m$$

be the unique sequence of subspaces of $\mathbf{C}^m$ satisfying 16.7, and let the integers k_i, m_i and ℓ_{ij} be defined as at 16.7. Then the stratum of the $GL(m)$ stratification of X to which $(L_1,\ldots,L_r)$ belongs is labelled by the vector

$$\beta = (k_1/m_1,\ldots,k_1/m_1,k_2/m_2,\ldots,k_s/m_s) \in \mathbf{t}_+$$

in which k_i/m_i appears m_i consecutive times for each i.

Proof. Denote the sequence $(L_1,\ldots,L_r)$ by x. We wish to show that $x \in S_\beta$.

Let $e_1,\ldots,e_m$ be the standard basis of $\mathbf{C}^m$. Then the vector space $\bigotimes_{1 \leq j \leq r} \Lambda^{\ell_j} \mathbf{C}^m$ has a basis consisting of elements

$$E_t = \bigotimes_{1 \leq j \leq r} e_{t(j,1)} \wedge \ldots \wedge e_{t(j,\ell_j)}$$

where t runs over all sequences

$$t = \{\, t(j,i_j) \mid 1 \leq j \leq r,\ 1 \leq i_j \leq \ell_j \}$$

of integers satisfying $1 \leq t(j,1) < \ldots < t(j,\ell_j) \leq m$ for each j.

Each of the subspaces $L_j \subsetneq \mathbf{C}^m$ is spanned by the rows of some $m \times \ell_j$ matrix A_j. The embedding of X in $\mathbf{P}(\bigotimes_{1 \leq j \leq r} \Lambda_{\ell_j} \mathbf{C}^m)$ sends x to the point with homogeneous coordinates

$$x_t = \prod_j \det A_{j,t}$$

where $A_{j,t}$ is the submatrix of A_j consisting of the columns $t(j,1),\dots,t(j,\ell_j)$.

Assume that $x_t \neq 0$ for some t. Then for each j the columns $t(j,1),\dots,t(j,\ell_j)$ of A_j are independent. Replacing x by gx for some appropriate $g \in GL(m)$ we may assume that M_i is spanned by the basis vectors $\{e_k \mid k \leq \dim M_i\}$ for $1 \leq i \leq s$. Then it is easy to check that

$$\dim (M_i \cap L_j) \leq \# \ \{i_j \mid 1 \leq i_j \leq \ell_j,\ t(j,i_j) \leq \dim M_i\}$$

for each i and j. By the definition of the integers k_i it follows that

$$(^*) \qquad k_1 + \dots + k_i = \sum_j \dim(M_i \cap L_j) \ \leq \ \# \{(j,i_j) \mid t(j,i_j) \leq \dim M_i\}$$

for $1 \leq i \leq s$.

The torus T_c acts diagonally on $\otimes_j \Lambda^{\ell_j} C^m$ with respect to the basis consisting of the elements E_t. For each $1 \leq k \leq m$ let χ_k be the character of T_c given by

$$\mathrm{diag}(a_1,\dots,a_m) \to a_k$$

and identify characters of T_c with elements of $t^* \cong t$ in the usual way. Then T_c acts on E_t as multiplication by the character

$$\alpha_t = \sum_{1 \leq j \leq r} \Big(\sum_{1 \leq i_j \leq \ell_j} \chi_{t(j,i_j)} \Big)$$

Therefore as $m_i = \dim M_i/M_{i-1}$ it follows from the definition of β that

$$\alpha_t.\beta = \sum_{1\leq i\leq s} (k_i/m_i) \ \# \{(j,i_j) \mid \dim M_{i-1} \leq t(j,i_j) \leq \dim M_i\}.$$

Here . denotes the standard inner product. The inequality (*) above implies that there are at least k_1 distinct choices of (j,i_j) such that $t(j,i_j) \leq \dim M_1$, at least k_2 more such that $t(j,i_j) \leq \dim M_2$ and so on. Since $k_1/m_1 > \ldots > k_s/m_s$ it follows that

$$\alpha_t.\beta \geq k_1^2/m_1 + \ldots + k_s^2/m_s = \|\beta\|^2 .$$

Moreover $\alpha_t.\beta = \|\beta\|^2$ if and only if

$$k_i = \# \{(j,i_j) \mid \dim M_{i-1} < t(j,i_j) \leq \dim M_i)$$

for $1 \leq i \leq s$.

By 16.7 L_{ij} is the image of L_j in M_i/M_{i-1} for each i and j, and by assumption $M_i = W_1 \oplus \ldots \oplus W_i$ where W_i is spanned by the basis vectors $\{e_k \mid \dim M_{i-1} \leq k \leq \dim M_i\}$ for each i. So we can identify L_{ij} with the orthogonal projection of $L_j \cap M_i$ onto W_i. We then see by looking at the matrices A_j that the coordinates x_t such that t satisfies $\alpha_t.\beta = \|\beta\|^2$ are the same as the coordinates of

$$(L_{11},\ldots,L_{sr}) \in \prod_{i,j} G(\ell_{ij}, M_i/M_{i-1})$$

when this product is embedded in projective space. In particular $x_t \neq 0$ for some t such that $\alpha_t.\beta = \|\beta\|^2$. Since we know that $x_t = 0$ when $\alpha_t.\beta < |\beta|^2$ this tells us that x lies in the subset Y_β of X as defined at 12.18 (or at 4.6).

It is now easy to check from either definition 12.18 or definition 4.6 that

$$p_\beta(x) = (L_1',\ldots,L_r')$$

where $L_j' = L_{1j} \oplus \ldots \oplus L_{sj}$ for $1 \leq j \leq r$. Also the component of Z_β containing $p_\beta(x)$ is isomorphic to the product

$$\prod_{i,j} G(\ell_{ij}, M_i/M_{i-1}),$$

and the stabiliser $\mathrm{Stab}\beta$ of β can be identified with

$$\prod_i GL(M_i/M_{i-1})$$

acting on this product in the obvious way.

It remains to check that $p_\beta(x) \in Z_\beta^{ss}$, for which it suffices (see definition 12.20) to show that

$$m(p_\beta(x);\lambda) \leq \lambda.\beta$$

for every $\lambda \in M(\mathrm{Stab}\beta)$. Alternatively from the symplectic viewpoint it is enough to prove that

$$\beta \in \mu(\overline{(\mathrm{Stab}\beta)\, p_\beta(x)})$$

(see 6.18).

By 16.7(a) each sequence $(L_{i1},\ldots,L_{ir})$ is semistable for the action of $SL(M_i/M_{i-1}) \cong SL(m_i)$ on

$$\prod_j G(\ell_{ij}, M_i/M_{i-1}).$$

Hence $m((L_{i1},\ldots,L_{ir});\lambda) \leq 0$ for all $\lambda \in M(SL(m_i))$, which implies that

$$m(L_{i1},\ldots,L_{ir};\lambda) \leq \lambda.(k_i/m_i,\ldots,k_i/m_i)$$

for all $\lambda \in M(GL(m_i))$ (compare 6.4). Since $p_\beta(x)$ may be identified with

$$(L_{11},\ldots,L_{sr}) \in \prod_{i,j} G(\ell_{ij}, M_i/M_{i-1})$$

it follows easily that

$$m(p_\beta(x);\lambda) \leq \lambda.\beta$$

for all $\lambda \in M(\prod_i GL(m_i))$. Hence $p_\beta(x) \in Z_\beta^{ss}$ so $x \in Y_\beta^{ss}$. There is an analogous argument using symplectic methods.

This completes the proof of proposition 16.9.

Note that for convenience in 16.9 we worked with $GL(m)$ not $SL(m)$. However when considering quotients it is better to work with $SL(m)$. For the central one-parameter subgroup of $GL(m)$ acts trivially on X and makes every point of X unstable for $GL(m)$.

Proposition 16.9 gives us an inductive formula for the equivariant Betti numbers of the semistable stratum in

$$X = \prod_{1 \leq j \leq r} G(\ell_j, C^m)$$

under the action of $SL(m)$. It is

16.10 $$P_t^{SL(m)}(X^{ss}) = P_t(X)\,P_t(BSL(m))$$

$$- \sum_{\beta,\underline{\ell}} (1-t^2)^{1-s}\, t^{d(\beta)} \prod_{1\le i\le s} P_t^{SL(m_i)}\Big(\big(\prod_{1\le j\le r} G(\ell_{ij}, m_i)\big)^{ss}\Big).$$

The sum is over all vectors $\beta \in t_+$ and sequences

$$\underline{\ell} = \{\ell_{ij} \mid 1 \le i \le s,\ 1 \le j \le r\}$$

such that there are integers $k_i \ge 0$ and $m_i > 0$ satisfying

$$k_1/m_1 > \dots > k_s/m_s,$$

$$\sum_i m_i = m, \quad \sum_j \ell_{ij} = k_i, \quad \sum_i \ell_{ij} = \ell_j$$

and

$$\beta = (k_1/m_1, \dots, k_1/m_1, k_2/m_2, \dots, k_s/m_s)$$

with each k_i/m_i appearing m_i times. Also

$$d(\beta) = \sum_{1\le i<j\le s} 2(k_i - m_i)\, m_j.$$

The factor $(1 - t^2)^{1-s}$ appears because in 16.9 we worked with $GL(m)$ not $SL(m)$.

16.11. Remark. In this example it is possible to show that the stratification is equivariantly perfect for any field of coefficients, not just the rationals **Q**. The proof is essentially the same as for **Q**. It works for all fields because $GL(m)$ is torsion-free, and because it is possible to find for each $\beta \in B$ a subtorus T_β which fixes Z_β pointwise and whose action on $T_x S_\beta$ is $\mathbf{Z}$-primitive, not just $\mathbf{Q}$-primitive, for each $x \in Z_\beta$ (cf. the proof of 5.4 and [A & B] theorem 7.14). We deduce that the GL(m)-equivariant cohomology of the semistable stratum has no torsion. Since PGL(m) acts freely on the semistable points, it follows from considering spectral sequences that the quotient variety has p-torsion for the same primes p as PGL(m), that is for $p \leq m$.

16.12. Ordered points in a projective plane

As a special case of the last example consider the diagonal action of SL(3) on $(\mathbf{P}_2)^n$. The first value of n for which the quotient is interesting is $n = 5$. Then $3 \nmid n$ so by 16.6 the stabiliser of every semistable point is finite.

Suppose $x \in X = (\mathbf{P}_2)^5$. By 16.5 x is semistable if no point of $\mathbf{P}_2$ occurs in x with multiplicity greater than $n/3$ and no line contains more than $2n/3$ components of x. If a point occurs with multiplicity $k > n/3$ and no line in $\mathbf{P}_2$ contains more than $2k$ components then x lies in the stratum labelled for GL(3) by

$$\beta = (k, (5-k)/2, (5-k)/2)$$

(see 16.9). If a line contains $k \geq 2n/3$ components then either a point of this line occurs with multiplicity $k_1 > k/2$ so that

$$\beta = (k_1, k-k_1, 5-k),$$

or else no such point occurs and

$$\beta = (k/2,\ k/2, 5-k).$$

So the stratification is given by the following table. The indices β here are indices for the GL(3)-stratification; the indices for SL(3) are given by replacing each β by $\beta - (5/3,5/3,5/3)$.

β	points x lying in S_β	contribution to $P_t^{SL(3)}(X)$
(5/3,5/3,5/3)	semistable for SL(3)	$P_t^{SL(3)}(X^{ss})$
(5,0,0)	all components coincide	$t^{16}(1-t^2)^{-1}(1-t^4)^{-1}$
(4,l,0)	4 components coincide	$5t^{12}(1-t^2)^{-2}$
(3,1,1)	3 components coincide, others linearly independent	$10t^8(1-t^2)^{-2}$
(5/2,5/2,0)	all components lie in a line, at most 2 coincide	$t^6(1-t^2)^{-1}(1+5t^2+t^4)$
(3,2,0)	all components lie in a line, at most 3 coincide	$10t^{10}(1-t^2)^{-2}$
(2,2,1)	4 components lie in a line, at most 2 coincide	$5t^4(1-t^2)^{-2}(1+3t^2-t^4.)$
(2,3/2,3/2)	2 components coincide, no 4 lie in a line	$10t^4(1-t^2)^{-1}$

By applying 16.10 we obtain the Betti numbers of the quotient $M = X^{ss}/SL(3)$. The Poincaré series of M is

$$P_t(M) = (1+t^2+t^4)^5(1-t^4)^{-1}(1-t^6)^{-1} - t^{16}(1-t^2)^{-1}(1-t^4)^{-1}$$
$$- (1-t^2)^{-2}\{5t^{12}+10t^8+10t^{10}+5t^4(1+3t^2-t^4)\}$$
$$- (1-t^2)^{-1}\{t^6+5t^8+t^{10}+10t^4\}$$

which works out as $1 + 5t^2 + t^4$. Here the inductive formula 16.10 was used rather than an explicit formula involving β-sequences. The former was quicker because the Poincaré series $P_t^{SL(2)}(((\mathbf{P}_1)^n)^{ss})$ have already been calculated.

When $n = 6$ we are no longer in a good case, and the series $P_t^{SL(3)}(X^{ss})$ is not a polynomial. When $n = 7$ we get

$$P_t(M) = 1 + 7t^2 + 29t^4 + 64t^6 + 29t^8 + 7t^{10} + t^{12} .$$

In general one finds that if $3 \nmid n$ the Betti numbers of the quotient M for the action of $SL(3)$ on $(\mathbf{P}_2)^n$ are given by

$$b_{2j} = a_j + 2a_{j-1} + \dots + (j+1)a_0$$

for $0 \leq j \leq 2(n-5)$, where a_d is given by the formula

$$\sum_{0 \leq b \leq d/2} \frac{(n-2)!}{(d-2b)!b!(n-2-d+b)!}\{(b+1-\frac{n(n-1)}{b+1}\left[\frac{\chi_1(b)-\chi_2(b)}{n-b-1} + \frac{\chi_3(b)}{n-d-b-1}\right]\}$$

$$- \sum_{n/3 \leq k \leq n} \frac{n!(\chi_4(k) - \chi_5(k))}{k(n-k)!(d-n+k+2)!(d+1)!}$$

and $\chi_1, \chi_2, \chi_3, \chi_4, \chi_5$ are the characteristic functions of the intervals

$[\max(n/3,d-n),d/2]$,

$[n/3,\min(d/2,(2d-n+1)/3)]$,

$[\max(d+1-n,(d-1)/3), \min(d+1-2n/3,d/2-1]$,

$[n-d-2,2(n-d-2)]$,

$[2(n-d-2),n]$.

References

[A & M] R. Abraham and J. Marsden, Foundations of mechanics, Benjamin, 1978.

[Ar] V.I. Arnold, Méthodes mathématiques de la mécanique classique, Editions de Moscou, 1976.

[A1] M.F. Atiyah et al., Representation theory of Lie groups, London Math. Soc. Lecture Note Series 34, 1979.

[A2] M.F. Atiyah, Convexity and commuting Hamiltonians, Bull. London Math. Soc. 14, 1-15 (1982).

[A & B] M.F. Atiyah and R. Bott, The Yang-Mills equations over Riemann surfaces, to appear in Phil. Trans. Royal Soc. London, A 308, 523-615 (1982).

[B] A. Borel, Linear algebraic groups, Benjamin, New York, 1969.

[B-B] A. Bialynicki-Birula, Some theorems on actions of algebraic groups, Annals of Math. 98, 480-497 (1973).

[B-B&S] A. Bialynicki-Birula and A.J. Sommese, Quotients by C^* and SL(2,C) actions, to appear in Trans. Amer. Math. Soc.

[Bi] D. Birkes, Orbits of linear algebraic groups, Annals of Math. 93, 459-475 (1971).

[Bl] A. Blanchard, Sur les variétés analytiques complexes, Ann. Sci. Ecole Norm. Sup. 73, 178-201 (1956).

[Bo] R. Bott, Nondegenerate critical manifolds, Annals of Math. 60, 248-261 (1954).

[C & G] J.B. Carrell and R.M. Goresky, A decomposition theorem for the integral homology of a variety, preprint.

[C & S] J.B. Carrell and A.J. Sommese, Some topological aspects of C^* actions on compact Kähler manifolds, Comment. Math. Helvetici 54, 567-582 (1979).

[D] A. Dold, Lectures on algebraic topology, Springer-Verlag Berlin-Heidelberg, 1972.

[D1] P. Deligne, Théorie de Hodge I, Actes du Congrès International des Mathématiciens, Nice, 1970, II, Publ. Math. IHES 40, 1971, and III, Publ. Math. IHES 44, 1974.

[D2] P. Deligne, Poids dans la cohomologie des variétés algébriques, Actes du Congrès International des Mathématiciens, Vancouver, 1974.

[G & H] P. Griffiths and J. Harris, Principles of algebraic geometry, Wiley, New York, 1978.

[G & S] V. Guillemin and S. Sternberg, Convexity properties of the moment mapping, Invent. Math. 67, 491-513 (1982).

[H] P. Hartman, Ordinary differential equations, Wiley, New York, 1964.

[Ha] R. Hartshorne, Algebraic geometry, Springer-Verlag, New York, 1977.

[He] S. Helgason, Differential geometry and symmetric spaces, Academic Press, New York, 1962.

[Hes] W.H. Hesselink, Uniform instability in reductive groups, J. Reine Angew. Math. 304, 74-96 (1978).

[H & N] G. Harder and M.S. Narasimhan, On the cohomology groups of moduli spaces of vector bundles on curves, Math. Annalen 212, 215-248 (1975).

[K] G. Kempf, Instability in invariant theory, Annals of Math. 108, 299-316 (1978).

[K & N] G. Kempf and L. Ness, The length of vectors in representation spaces, Springer Lecture Notes 732, 233-242 (1978).

[Ki] F.C. Kirwan, Sur la cohomologie des espaces quotients, C.R. Acad. Sci. Paris, 295, Serie I, 261-264 (1982).

[Ki2] F.C. Kirwan, Desingularisations of quotients of nonsingular varieties and their Betti numbers, preprint.

[Ki3] F.C. Kirwan, On spaces of maps from Riemann surfaces to Grassmannians and applications to the cohomology of moduli of vector bundles, preprint.

[L] S. Lang, Algebraic groups over finite fields, Amer. J. Math. 78, 555-563 (1956).

[M] D. Mumford, Geometric invariant theory, Springer-Verlag, New York, 1965. 2nd edition, D. Mumford and J. Fogarty, Springer-Verlag, Berlin-New York, 1982.

[Mi] J. Milnor, Morse theory, Annals of Math. Studies, Princeton University Press, 1969.

[M & W] J. Marsden and A. Weinstein, Reduction of symplectic manifolds with symmetry, Reports on Math. Phys. 5, 121-130 (1974).

[N] P.E. Newstead, Introduction to moduli problems and orbit spaces, Tata Institute Lectures 51, Springer-Verlag, Heidelberg, 1978.

[Ne] L. Ness, A stratification of the null cone via the moment map, to appear in Amer. J. Math.

Library of Congress Cataloging in Publication Data

Kirwan, Frances Clare, 1959-
Cohomology of quotients in symplectic and algebraic geometry.

(Mathematical notes ; 31)
Bibliography: p.
1. Group schemes (Mathematics) 2. Algebraic varieties. 3. Homology theory. 4. Symplectic manifolds. I. Title. II. Series: Mathematical notes (Princeton University Press) ; 31.
QA564.K53 1984 512'.33 84-15143
ISBN 0-691-08370-3 (pbk.)

Frances Clare Kirwan is a Junior Fellow at the Society of Fellows at Harvard University.

Milton Keynes UK
Ingram Content Group UK Ltd.
UKHW051130310824
447656UK00017B/230

9 780691 083704